PIED PIPERS IN NORTH-EAST INDIA

PIED PIPERS IN NORTH-EAST INDIA

Bamboo-flowers, Rat-famine and the Politics of Philanthropy (1881–2007)

SAJAL NAG

MANOHAR
2008

First published 2008

ISBN 81-7304-311-6

Published by
Ajay Kumar Jain for
Manohar Publishers & Distributors
4753/23 Ansari Road, Daryaganj
New Delhi 110 002

Typeset by
Kohli Print
Delhi 110 051

Printed at
Lordson Publishers Pvt. Ltd.
New Delhi 110 007

To my friend
Dr. M. SATISH KUMAR
who deserves much more

'We've Chivvied the Naga and Looshai
We have given the Afridiman Fits'

RUDYARD KIPLING (1890)

Contents

Preface

In the beginning of 2001 the elders of the Mizo tribe of Mizoram forecasted that a famine was likely to strike the state by the year 2007. It would be a rat-famine induced by the reproductive flowering of the bamboo plants that carpet the hills of the region. True to the prediction there were reports of bamboo flowering by the middle of the year. Reporters went to the jungles and before long photographs of bamboo plants with their flowers appeared in the news media. Reports indicated that there was a visible increase in the jungle rat population. As the news spread, there was a massive discussion among various circles in north-east India about the impending famine resulting from bamboo-flowering in Mizoram and the consequent increase in the rodent population who feed on human food. It indeed appeared that the famine was likely to occur in 2007. A similar phenomenon in 1959 sparked a twenty-year long insurgency which ended only in 1986 after a traumatic experience for both the people of Mizoram as well as the Indian state. Fearing a repeat of a similar situation both the state and central governments were not leaving anything to chance and doing everything possible to pre-empt the event. The whole matter was intriguing.

I recalled that during my search in the Mizoram State Archives for material relating to a different subject, I had accidentally come across information on bamboo-flowering in the same region during the colonial period. On the basis of that material I wrote a paper for the *Cambridge Journal of Environment and History* some time ago. As interest grew on the subject I used the material and published another paper in *Economic and Political Weekly* (December 2001) connecting past developments to the present. Reading the paper, Professor Shiv Vishwanathan of Centre for Studies in Developing Societies, New Delhi asked me if I would write a small monograph on the issue of bamboo-flowering and rat-famine in Mizoram if he provided the finances. I was happy to write but was not sure that there was sufficient material for writing a book. But as he egged me on, I ventured into the project. The credit, for starting the project towards this monograph should go entirely to Shiv and his colleague Chandrika Singh. I am immensely thankful to them for that. Although I prepared a preliminary report on my research, despite encouragement from both Shiv and Chandrika, I held back its publication as I realized there was still a lot of material available on the subject both in India and abroad. I had to wait for the necessary finances to be able to consult the materials. The opportunity came when I was awarded the Commonwealth Fellowship by the Association of Commonwealth Universities, United Kingdom, to pursue my research on the subject. Professor Julian Orford, Chairman of the School of Geography, Queen's University, Belfast, Northern Ireland, offered me affiliation to his university as he found the subject matter 'intriguing'. Dr Satish Kumar, who is a childhood friend from our Shillong days and co-incidentally based in the same school now, agreed to supervise the work. Besides academic association, the tenure provided an opportunity to relive our childhood. Professor David Livingstone has shown an unfailing interest in my work. So did others in the school who were instrumental in making my stay in Belfast homely and memorable. David Zou was my constant companion in Belfast while Suresh Rohilla and his wife Padma, daughter Jia were people whom I turned to whenever I had any difficulty in a foreign land. My cousin Raju rosted me in London with the best of care. No thanks are enough to acknowledge my regards and gratitude to all of them.

Thanks are due to the archivists and staff of Mizoram State Archives, Aizawl; District Record Room, Silchar; State Archives, Dispur; National Archives, New Delhi; British Library, London; Centre for South Asian Studies Archives, Cambridge University; Librarian and Staff, Queen's University, Belfast. I also would like to thank Dr O.P. Kejariwal, former Director, Nehru Memorial Museum & Library, New Delhi, and his wife Debolina Banerjee-Kejariwal for adopting me as a member of their family. They are people who I always look up to for sustenance. My friends

Madhusudan Karmakar and Muhammed Parwez with their respective spouses have always been supportive. Thanks are also due to Professor David Reid Syiemlieh and Imdad Hussein of the Department of History, North Eastern Hill University, Shillong, for their interest in my research and help through archival material. Suryasikha Pathak have helped me with photographs of bamboo flower and fruit. Thanks are also due to Professor Willem von Schandel for permitting me to use a photograph from his book. My daughter Tuntuni, despite missing me encouraged me to do what I was doing. My wife, Tejimala had to look after our daughter all-alone along with her regular university duties while I was away in Belfast. My parents, brothers, sister and their spouses, sisters-in-law, Subala and Binu attended her when she had met with a serious car accident requiring a major surgery while I was away. My friends Shyamananda, Abhijit da, Sudarshan, Raju and Shipra di were there in time of that crisis. I cannot thank any of them enough. None of the above-mentioned are responsible for the errors, if any, either of facts or interpretation of the material presented in the book. Those are all mine.

SAJAL NAG

Map 1: North-East India.

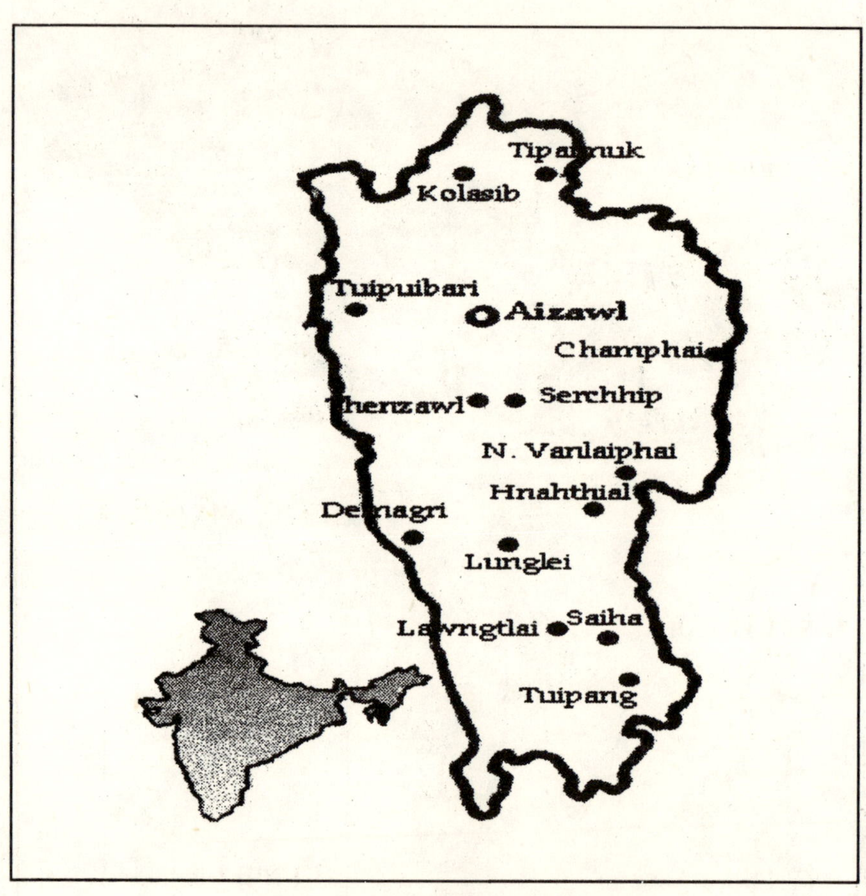

Map 2: Mizoram State of India.

CHAPTER 1

Introduction

The history of mankind is a story of man's ceaseless struggle against nature. This is because nature, which facilitated men's survival, often was the cause of his extinction as well. In this struggle only the fittest survived. Nature often helped the civilizations flourish but at other times it created adversities to its detriment. Natural hostility could be of two broad types: one, an environment unfavourable to human survival and two, sudden and unforeseen devastation by nature in the form of calamities. The calamities too could hit in myriad ways: earthquakes, floods, volcanic eruptions, tsunamis, landslides, avalanches, cyclones, hurricanes, twisters, epidemics, typhoons, storms, tornadoes, draught and famines.[1] Human history is therefore a story of challenge and responses to nature. Civilization flourished where mankind was able to respond to these challenges effectively and it perished wherever it failed.

History cannot exclude this aspect of man's struggle against nature from its purview. In fact, traditional historiography has always included calamities and its impact on man as a part of its study. The Bible has talked about the Great Flood; in the Middle Ages Europe witnessed the Black Death and the bubonic plague. Famines played a very important role in shaping the history of England and Ireland. It is only in recent times with the Rankean revolution in Germany which emphasized only political history, the study of calamity and man's day-to-day struggle for survival has been relegated into the background. But this historiographical neglect cannot minimize the importance of natural calamities in the making or breaking of civilizations.

Primitive man took shelter in the safety of caves and mountain cavities to escape the vagaries of nature. But since then men have made tremendous progress in science and technology.[2] But he is still as helpless against natural calamities. Earthquakes in the seabed and the consequent eruption of sea water are part of the life of coastal regions. During the tsunami of December 2004 about 2,50,000 people perished. Frequent earthquakes and volcanic eruptions are an integral part of Japanese life. The Japanese have succeeded in minimizing the effects of these natural calamities on them as they developed early warning knowledge systems and survival tactics during such crises. In fact, it is believed that the devastation by natural calamities has increased manifold and intensified in recent times. There was a devastating earthquake in northern Iran, volcanic eruptions in Philippines and Alaska, a deluge in Texas and Bangladesh, cholera epidemic in Peru, widespread drought in California and East Africa, Hurricane Andrews, Katrina and Rita in the United States. Hurricane Erick had destroyed the island of Hawaii. India too has had its share, a ruinous typhoon in Orissa coast (1999), the earthquakes in Latur (1993), Bhuj, Gujarat (2002), and the two sides of Kashmir in India and Pakistan (2005), frequent floods in West Bengal, Bihar, Assam and Maharashtra in recent times. Research on the nature and causes of natural calamities have revealed that since the Second World War up to 1986 the intensity and frequency of natural calamities have actually increased over the last 50 years.[3] The report further stated that though the number of major natural calamities has not increased in number, the frequency of smaller calamities have increased manifold in recent times. Although even the most advanced countries were not found to be free from the vagaries of nature, what is alarming is that the developing world and formerly colonized countries were seen to be more vulnerable to such destructions. These regions also reported more devastation and casualties. The research conducted by the United Nations has also shown similar results. During 1987–9 the greatest number of natural calamities took place in countries other than North America. Out of these more than half took place in developing countries affecting a massive number of people.[4] Analysts feel one reason is the tremendous increase in global population since the First World War. The developing countries have experienced population explosions. Livelihood requirements have compelled people to fan out and settle in hitherto uninhabited sites like the river basins, ports, sea shores, mountains and volcanic foothills. Such sites are dangerous as they are close to calamity sites with few escape routes rendering them most vulnerable. Since nature is unpredictable and warnings have often proved ineffective, loss of life and property have multiplied. Moreover due to the development of science and technology people have forgotten their traditional coping mechanisms and survival strategies *vis-à-vis* natural calamites and have increasingly become dependent on the state and its agencies for succour during these crises. The over-dependence on the state has actually rendered people more vulnerable as the state is bureaucratic in nature and slow and too cautious in acting. Moreover, a state has its own perceptions and theories of calamities and disasters. What is a calamity to the affected people may not be in the scheduled list of calamities of a particular state. All natural calamities are not 'natural' or 'calamities' at all as far as the state is concerned. Even if a natural calamity hits people in a terrible manner, the state may not come to the rescue of its people when it does not fit the listed description of a natural calamity. Again some calamities are defined as 'national' in character while others are 'regional' according to the policies of a state.[5] In such cases the nature of relief and rehabilitation could vary according to the scale befitting a national or regional event. It is obvious that the definition of 'natural', 'national' and 'regional' depends on the strength of the respective lobbies and the power of the political elite.

In a number of cases the state itself is responsible for causing 'natural' calamities. In North America a number of massive dams were constructed during the nineteenth century upsetting the geological formations of the region. In India the state has gone ahead with the construction massive dams on the Narmada and Tehri rivers despite the fact that they fall in a seismically vulnerable region and are submerging huge areas of human habitat, drowning peoples' life and property earned through the generations and also threatening the region with massive earthquakes. Natural calamities have not always been an event of fear. It is found that only in modern times natural calamities and disasters have become a prospect of fear. But traditionally people's values were different and men had learnt to fight adversities of nature and evolved coping mechanisms to deal with such events. Thus, despite the devastation it caused, neither nature nor its adversities was feared. With mankinds increasing use of technology in securing mastery over nature, they became complacent and arrogant. But whenever nature proved more powerful than men thought it to be, fear developed in greater proportions. Calamities thus changed the values and perception of human beings. Natural disasters are also no more considered as solely natural.[6] It is intricately connected with the network of balance between nature, social, political, economic and technology. Hence natural disasters are not absolutely 'natural' in modern times. Experts felt that the vulnerability of a region that has suffered from a natural calamity cannot be studied without 'contextualizing' them from the larger issues of environment, social and political structure.[7] In fact modern-day political practices have appropriated natural calamities. Calamities have been used by men to acquire colonies, subdue people, annex territories, secure power, infuse ideologies, transform religion, induce culture and instigate rebellion. Calamities have thus had an all-pervading presence in modern history.

It is believed that the increased devastation wrought by nature on human beings has been a result of men's own interference with nature, thereby upsetting its delicate balance. Rivers have always inundated its basin. Therefore 'inundation' has traditionally been a geographical term. Nile, Hoang-ho, Amazon, Tigris, Euphrates, Indus and Ganges are rivers that have been the cradle of most advanced of civilizations. The water of these rivers often spilled over during the rainy season inundating its basins but seldom did they affect the human habitat adversely. This was because geography had evolved their own channel network through which the excess water flowed back to the rivers. Colonial regimes were the first to tinker and interfere with the rivers, constructing huge dams on it. They also exploited the forest resources denuding it to the extent that it was no longer able to absorb the excess water. The extreme emphasis on urbanization as a form of modern civilization without any consideration to the sewage system let the excess water enter human habitat and cause displacement and devastation. The geographical term 'inundation' was given a new name by colonial regimes—floods. The indigenous people witnessed the transformation of life giving rivers into instruments of devastation. Thus the nationalist critics of colonialism for example preferred to call floods in colonial Assam as 'deluge'. In fact, the last decades of the last century could as well be described as the decades of natural calamities. Hence, 'it would be a great mistake, in our time, to ignore the existence and role of such events in Earth's history. . .'.[8]

Humankind have always had to cope with calamities. Such catastrophes have often made, unmade and greatly influenced the shaping of human civilization. Thus, one may say calamities have formed an integral part of human history. There have been many such communities, nations and civilizations who struggled against calamities for their survival. Some perished, others survived. In fact, calamities are inseparable from human history. At the same time human history is not a history of the destruction of human civilization by catastrophes but also mankind's ceaseless struggle to understand them and acquire a mastery over nature. As, with the passage of time, the emphasis of history has shifted from dynasties to the downtrodden, politics to people, events to environment; new historiographical schools have given shape to a new sub-discipline in history called Calamity Studies, wherein historians study the impact of calamities on human history.

The north-east of India has immense potential for Calamity Studies. The major calamities that are frequent in north-east India are diseases like kala-azar, malaria, small pox, natural disasters like floods and earthquakes, environmental catastrophes like landslides, famine and droughts. In Assam the popular proverb is that *jui, pani, yuin* (fire, water and white ants) are the three major agents of mass devastation in Assam. Floods of different intensity were almost an annual occurrence during the rainy season. There were recorded floods in 1240, 1570, 1642, 1786, 1781, 1825, 1842, 1862, 1867 and 1870 in Assam. Blights, floods and drought adversely affected the harvest of the peasants quite regularly.[9] The districts of Lakhimpur and Kamrup were prone to floods. The district of Sibsagar experienced severe drought in 1857, 1870 and 1872. There were several cases of fire reported throughout the nineteenth century in Assam. Another menace in Assam was the white ant invasion. John McCosh reported that

> white ants occupy a prominent place in the annual economy of Assam. In no parts of India their ravages are more destructive, they devour the very house they stand, from main posts sunk seven feet underground, to the last bundles of thatch upon the ridge. . . . The furniture required to be constantly looked after, the feet of a table or a chair may be eaten up though no outwards signs be discernible.[10]

These calamities which devastated the region repeatedly were not all 'natural'. For example, famines in Manipur from 1930s onwards were artificial and resulting from man-made shortages of essential food items.

As a calamity, famines have played a very important role in the political developments in north-east India. It is now well known that the 1959 bamboo-famine had sparked off the twenty-year long insurgency in the Mizo Hills. But what is not yet known is that anti-India sentiment in Manipur and Nagaland also developed as a result of widespread famine. The Naga Hills were reeling under severe food crisis during 1946–7 which neither the colonial government nor the Indian leadership cared about.[11] In Manipur, it was the Marwari and Sikh merchants who entered Manipur along with the British and in no time monopolized the trade and commerce of the tiny princely state. With only one-third of its land arable, the state produced just enough for its subsistence. But during the Second World War, to satisfy wartime needs, these merchants bought up all the rice produced in the valley and supplied them to the Allied army. It often created severe food shortages for local consumption thereby resulting in an artificial famine situation in Manipur. The sheer severity of the situation prompted the womenfolk of Manipur to organize an all-women movement protesting against the export of local rice by outside traders (called *mayangs* by the Meiteis). The movement is widely known as *Nupilan* (womens') movement, which broke out twice; once in 1904 and next in 1939–40. The breaking out of the *Nupilan* on the eve of Indian Independence had far-reaching implications as far as the shaping of anti-India attitude of the Meiteis was concerned. The artificial famine resulting from the hoarding of essential commodities by the Marwari and Sikh traders created the image of Indian traders as 'unscrupulous people'.[12] The continuation of such practice by these merchants despite protest movements and social ostracization only promoted the negative image of northern Indians in the Meitei mind. The success of the *Nupilan* movement only reinforced it.[13] In Mizo Hills the bamboo flowering and the consequent rat-famine (1957) sparked off the twenty years long bloody secessionist insurgency (1966–86). The tribal elders of Mizoram had warned the state government of Assam about an impending catastrophe in the form of bamboo flowering and consequent rat-famine. But Government of Assam dismissed the prediction as tribal superstition and ignored the warning. True to the prediction, the bamboo plants blossomed all over the hill ranges leading to an increase in the rat population who fed on human food. The state government was caught unawares and before they could take any ameliorative action, scores of lives were lost to starvation. The famine was over in time but the antipathy shown by the state to the Mizo tribals remained.[14] They demanded secession from India on the ground that they would not like to be part of a nation which did not come to their rescue in times of dire need. The movement graduated to organized insurgency, which the Indian state found difficult to tackle for a long time before the Mizos negotiated for an amicable settlement.

While political activity and social movements were organized around calamities like famine in colonial times, disasters like flood and landslides have been the pivot of politics in contemporary north-east India. Calamities have been used to highlight the neglect of the centre towards its north-eastern 'periphony'. Unfortunately there has not been much work on the impact these calamities have had on the people of the region. The objective of this monograph is to pioneer a study on the history and politics around the bamboo famine in the Mizo Hills from the 1880s to the 1960s. This was in reality a rat-famine resulting from the reproductive flowering of the bamboo plants which is a rare occurrence.

The association of the Mizo with *mautam* or Mizoram as the land where the bamboo plant flowers is indeed historical. As the Irish were historically associated with potatoes and famines, the Mizos have been associated with bamboo flowering and rat-famines. Very few people knew that Mizoram is the theatre of such a rare environmental event. Of the first persons to publicize the state's association with this phenomenon, was J.D. Baveja, who in the tradition of the colonial bureaucrats wrote a monograph on the people he was administering titled *Mizoram: The Land Where Bamboo Flowers*. From the title itself it is evident that Baveja had no prior knowledge about the phenomenon. When he learnt about it he was equally amazed by the mysteriousness of the entire event and therefore used it profitably as the title of his monograph on the Mizo people. However, in the book there is hardly anything about bamboo flowering.[15] It is more an ethnographic narrative on the history and society of the Mizos. Although the book did not provide any new insight on the Mizo tribe, its title immortalized the association of the Mizo people with bamboo flowering.

Mizos are indeed inseparable from the phenomenon of bamboo flowering—*mautam* as the Mizos call it. Monographs, semi-historical and literary works on the Mizo Hills have marked Mizoram with bamboo flowering. Almost two decades after Baveja wrote his book, Vasanthi, a litterateur from another corner of the country (Tamil Nadu), wrote a popular novel depicting 'a human story written on the backdrop of the Mizo life' which was titled as *When Bamboo Blossoms*.[16] A similar novel was published in Hindi in 1997 with bamboo flowering as the backdrop. It was authored by Prakash Mishra and was titled *Jahan Baans Phulta Hai* (Where Bamboo Flowers). Subsequently another work by E.J. Thomas detailing the socio-economic transformation, taking place in the state of Mizoram was published with the title, *Mizoram: Bamboo Hill Murmers Change*. Sociologist Shiv

Vishwanathan has a chapter on bamboo flowering in his book *A Carnival of Science*. The title of the chapter interestingly is 'House of Bamboo'. The present work will be an addition to this list. As can be seen later, bamboo famine has really chased the Mizos along their migration up to the present habitat. Not only this calamity kept their population dwindling, they would have perhaps shifted from their present habitat where they migrated only in the eighteenth century, had the British not halted their movement by aggression as well as helping them mitigate the calamity. The missionaries too taught them to take control of these recurring calamity before they could convince them to convert to Christianity in large scale.

The Phenomenon

Does bamboo flower too? This was the question my friends asked me when I told them about my present research. *Yes it does.* I replied with the confidence of the editor of an encyclopaedia. The barrage of questions that followed, I replied with equal élan. Little did they know that till I started my research I hardly knew anything about bamboo flowers and the connected rat-famine. I pored through volumes of encyclopaedia for study material to uncover these ecological mysteries. Botany is a strange subject. They do not address anything by their common names. They have their own names for them. Thus, Bamboo is called *Bambusoideae.* They also have their own mysteries about the world of plants, grass and trees. For example, I did not know that bamboo was not a tree; it is in fact a grass. It is strange because who would think that the grass in your lawn and the tall bamboo are of the same category? These bamboos also flower and bear fruits. Equally intriguing was the discovery that the reproductive flowering and fruition cycle takes place once in thirty, fifty and even hundred and twenty years. But there are also species which flower every year. These plants do not die but the variety which flower in longer spells begins to rot and then die after blooming. The subsequent generations of the plant are born out of these seeds but it takes three to four years. The jungle rats love the bamboo fruit containing seed to the extent that the entire rat population comes overground ignoring the risk of being killed by serpents, animals and human beings. It is also believed that the food actually enhances the reproductive health of the rodents to such an extent that they become prolific breeders thereby adding enormously to the rodent population. All these are basic knowledge to botanists and zoologists but a great enigma to the students of social sciences and humanities under the present curriculum in Indian universities. Thus a simple reproductive function of the bamboo plant is shrouded with so much mystery around it. But beyond the centres of power, knowledge was gathered through generations of experience and transmitted through oral traditions by people who live far beyond the boundaries of these knowledge metropolis. It is these tribal elders in the Indian state of Mizoram, who for generations, through their transmitted knowledge about bamboo-flowering and the organic rat–bamboo connection has correctly predicted the exact time and sequence of the next occurrence of this amazing ecological phenomenon. Sometimes the amazing mysteries of the earth are good to read about but no one actually sees it happening. It happens in distant lands, on the fringes of the mainstream. For instance, very few Indians, outside the state of Mizoram knew that such a mysterious occurrence actually takes place in one of the states of their own country. Even the populace in the neighbouring states in north-east India knew precious little about this unique environmental event. Within Mizoram itself, my interviews with Mizo youth, who were born after the last bamboo-flowering spell in 1977, showed that even they were no better informed. Yet amidst them, in the Mission Veng of the capital town of Aizawl, there is an 88-year-old octogenarian, D. Rokhuma,[17] who has been working consistenly for the past several decades to understand the phenomena and sensitize people about the event and its implications.

The flowering of bamboo is an unique phenomenon. Unlike most other living organisms, it flowers only once in its lifetime after which it dies. The other peculiarity of such flowering is the entire forest of bamboo plants flower at the same time, which is the reason that such an event is called 'gregarious flowering'.

Incidently, they die at the same time too. Records of bamboo flowering stretch back as far as ninth century AD when a particular variety of bamboo (*Phyllostachys bambusoides*) flowered in Japan.[18] In fact, it was the plantologist, Dai Kai Zhi who used the words *bamboo flowering* for the first time in his Manual of Bamboo.[19] Bamboos are arborescent grasses belonging to the family of *poaceae* and are grouped under the subfamily *bambusoideae.*[20] Out of the 110 general varieties and 1110–40 species of bamboo in the world, India, according to the latest reports accounts for 18 general and 128 species of bamboos. North-east India has extensive bamboo vegetation covering an area of 3.05 million ha. Out of this the state of Mizoram accounts for highest forest cover with bamboo. It has 9 general and 20 species. The reproductive cycle of bamboo vary from species to species. Based on the flowering cycle they are classified into three types:[21] (1) annual or continuous flowering (species which flower every year and do not die). (2) Gregarious or periodic flowering when the whole clump flower in an extensive area and die after seed setting. The flowering may continue for two or three years in an area or in the same clump. (3) Sporadic or irregular flowering which occurs in isolated clumps in one or to two in an area or in parts of one clump. Thus flowering periodicity of bamboos varies from three to one hundred and twenty years as may be seen from Table 1.1.

Scientifically the bamboo flowering phenomenon is called 'synchronic masting'.[22] This term is used by scientists to describe the 'synchronized production of seeds at long intervals by a population of plants'. Masting is the accurate term but flowering was the word coined by the colonial authorities, which gained acceptance in common vocabulary. For the colonials the flowering of bamboo was an incredible event which they recorded again and again but they never seemed to have looked for an explanation.[23] Among them were E.C.S. Baker,[24] H.R. Blanford,[25] E. Blatter,[26] J.W. Bradley,[27] D. Brandis[28] and I.F. Bourdillon[29] and in later times Father H. Santapu[30] and D. Chatterjee.[31] It was only in 1967 that the first synoptic survey was conducted by Daniel Janzen.[33] Janzen regretted that although natural flowering of bamboo was being seen as a unique event, the natural bamboo

TABLE 1.1: FLOWERING HABITS—PRIORITY SPECIES[32]

Species	Period between flowerings in years
1. *Bambusa balcooa*	35–45
2. *B. bamboo*	45–8
3. *B. blumeana*	45–100
4. *B. polymorpha*	35–60
5. *B. texilis*	Regularly found to flower, does not die after flowering
6. *B. tulda*	30–60
7. *B. vulgaris*	Rarely found to flower
8. *Cepha pstacjuum pergracile*	Sporadic
9. *Dendrocalamys asper*	Not known
10. *D. giganteus*	40–65
11. *D. latiflorus*	Sporadic
12. *D. strictus*	40–65
13. *Gigantochloa apus*	50–60
14. *G. Levis*	Sporadic, not fatal
15. *G. pseudoarundinacea*	50–60
16. *Guadua angustifolia*	Not known
17. *Melacanna baccifera*	35–6, 48, 60
18. *Ochlandra travancorica*	7 (and 23)
19. *Phyllostachys pubescens*	67 and sporadic
20. *P. bamusoides*	60 (and 120)
21. *Thyrsostachys siamensis*	35–50

forests were being denuded across the world.[34] During flowering, individual aerial stems sometimes live for much less time than their species cycles and flower only at the end of the cycle when an inborn signal initiates the formation of inflorescences. Fruit development in a few species has also been reported. The size and shape of bamboo fruits vary according to the species. The morphology of fruit was dependent on its character for identification of bamboos. Researches have furnished an account of bamboo fruits belonging to 17 general and 22 species. Although bamboo fruit are generally known as *Caryopsis*, based on morphology, researchers classified them into three types:

1. *Caryopsis*: The *pericarp* is membranous, thin, soft and adheres to the seed coat. The fruit has an apparent ventral suture, which is nearly as long as the whole fruit. An articulate navel is located at the fruit base.
2. *Glans* have hard, smooth, *crustacous pericarp*, separated from the seed coat. The fruit has no ventral sutra and navel.
3. *Bacca* has thick fleshy pericarp separated from the seed coat. It indicated that the morphology of starch grains can also be used as a distinguishing character for identification.

In certain parts of China, contiguous parts of Myanmar with India, Japan and South Africa, this phenomenon was known to occur. In Japan the *moso* species of bamboo bloomed, after 67 years in 1997.[35] The bloom of *moso* for the second time was actually checked in the Kamigamo Experimental Forest Station in July 1997. In Japan, there was a reported outbreak of Smith's red-backed mole (a small burrowing insect eating mammal of the family of *talpidae*) subsequent to the flowering of *sasa ishizichiana* at Mount Tsurngi-shikoku in 1954–5.[36] In southern China there was mass flowering of thorny bamboos in 1994, which coincided with severe floods, bringing immense suffering to the people.[37] In fact, the instance of bamboo flowering in China was brought to notice of the plant scientists in the 1980s by zoologists who were attracted by the plight of the Giant Pandas during the flowering of bamboo plants. It was then observed that there had been many instances of mass flowering of bamboos causing a complete loss of livelihood to local growers.[38] Their trust in the reliability of the resource base was shattered and in many instances the people moved away permanently from bamboo as their source of income. The Blue Mountain of Jamaica has one of the best-recorded instances of periodic flowering.[39] Around the same time such flowering was reported from Trinidad and Tobago as well. In South America too there was a reference of a bamboo bloom with corresponding increase in rodents in the second half of the 1800s.[40] In this case, a peak in rainfall accompanied bamboo flowering. Similarly, research done by Taylor Green of the Pest Infestation Control Laboratory, United Kingdom indicated that in East Africa, there was an excessive rodent population which followed heavy rainfall late in the season the next year.[41] The explanation given was that the late rainfall helped weed and grass to grow and survive longer than usual thereby providing protection to the rodents from predators, mainly birds and partly because these also provided additional food to the rodents. In Mizoram too, the 1977 bamboo flowering was accompanied by heavy rainfall.[42] In Thailand in the 1980s due to 'gregarious blooming' even the newly planted bamboos flowered and died. It is called 'gregarious bamboo flowering' because the bamboo clumps flower all at the same time in complete synchrony but only once in its lifetime. In the Balaghat district of the Central Provinces in 1869, the Kattang bamboo (*Bambusa arundinacae*) flowered and coincidentally there was a famine but no suggestion was made that there was any interrelation between the two occurrences.[43] The Indian states of Maharashtra and Uttarakhand also reported the flowering of bamboo but no famine or rat proliferation had been reported. There are other areas of India's north-east where these species of bamboo plants exists and the phenomenon also occurs. The famine did not take place since the affected area was not a human habitat. But, it is said that 'nowhere does the outbreak reach the proportion that it does in Mizoram'.[44]

The correlation of bamboo flowering with famine is interesting. Famines are simply food shortage. The shortage could be due to natural or artificial causes.[45] Among the natural causes, shortage caused by infestation of vermin has been listed as one.[46] But such famines were 'minor and localised'. Rats have been also listed as one of the vermin who through depredation cause food shortage. Rats eat almost anything that humans eat. They cause the most serious damage to the seeds of grain both before and after harvesting. Grain stored on farms is often not only eaten by rats but also rendered unsuitable for human consumption when mixed with rat dropping. With population explosion among the rats the destruction of foodstuff also increases. Therefore as far as foodstuff of human beings are concerned the rats have been identified as a major destroyer.[47]

The gregarious flowering of bamboo increases the food supply of the rodents. Shiv Vishwanathan described the situation as a carnival for the animals:

When the bamboo flowers it is the signal for a feast. Everyone loves the bamboo. Its fruits look like a yellowish-green pear with a tender, white kernel inside. Sometimes the flowering is in clumps, some time the masting takes place over 95000 square miles. . . . Animals love it. Birds migrate en masse. Janzen reports of a guinea fowl he found which had over 400 seeds in his stomach. Rats, Pigs and chicken ravage the area and eat till hunger pangs become a distant fantasy. It is gluttony on an enormous scale. Rabelais would have the right word for it not botanists. No Pied Piper could play a carnival song such as this. There have been reports of as many as 40–60 million rats migrating to a masting. All they do is eat and eat and reproduce. Nothing grows at this time for these rat populations remain in the area for two years. Rice and even root crops like yams and arum growing in the area are devastated. Often villagers have to abandon their homes. As the rats, pigs and birds eat their reproductive cycles shorten. It is a factory-farmer's dream as litter follows litter. For chickens—the pigs of the bird world—increased food consumption can lower the time of sexual maturation by fifty days. What is feast to nature, becomes a famine to man as rats, birds and pigs multiply.[48]

The 1977 *Thingtam* famine in Mizoram was a living testimony to this occurrence.

Subsequent to the flowering of bamboos rats multiplied in thousands. They moved from village to village from *jhum* plot to *jhum*, in groups of thousands. Often in a matter of hours whole acres of standing crop, mainly rice would be destroyed. Though efforts were made at baiting and fumigation, traps were laid and pits dug around the *jhums* the effect was marginal. The number of rats was so overwhelming that often pits would get filled in by the bodies of the front-line rats while those following behind would walk over the bodies and enter the fields.[49]

This phenomenon however seriously affects the normal balance of nature. Animals dependent on bamboo vegetative growth such as the Giant Panda and Mountain Gorilla loses a favoured food source entirely after a flowering episode.[50] A glut of bamboo fruit may also incite an explosion of population of rodents that eat the fruits. For example the flowering of *muli* or *terak* bamboo (*me locanna bambusoides*) in its native habitat around the Bay of Bengal in cycles of mostly 30–5 years leads to disaster. The accumulation of avocado-sized fruits promotes a rapid increase in rodent population, which lead to the loss of human food supplies and epidemic of rodent carrying disease.[51] A research at the Dehradun Institute of Forest Research on the connection between bamboo and rats found that the fruit contained:

1. Starch (on zero moisture basis)	50.240 %
2. Protein	11.556 %
3. Fat	0.231 %
4. Ash	3.030 %
5. Moisture	9.400 %
6. Others	26.493 %

It states that the rats feed on bamboo seeds which are rich in enzymes like *deamidase*, *amylase* and *amygdalins*. These trigger an overdose of oestrogen, enhancing fertility of the rodents enabling them to multiply at very rapid scale. According to a research of Dr. A.K. Ghose, a zoologist based in North Eastern Hill University, Shillong,[52] the fruit has high protein of nearly 12 per cent and very high starch content of about 50 per cent besides Vitamin A which helps augment the fertility of rodents.[53] No clear causal link between the flowering bamboos and the increase in rats has been established, though various theories exist. According to one theory[54] as bamboo seed contain a high concentration of Protein—nearly 11 per cent—perhaps by eating them the cannibalistic urges of the male rats toward the young is deadened. Also high intake of protein might strengthen the rats thus lowering the rate of infant mortality. It is also conjectured that some hormonal change took place in the rats due to the excessive protein that the bamboo fruit contained enabling the female rats to produce a litter much earlier than in normal circumstances. Even in normal circumstances however the rats are prolific breeders.[55] Although larger types reproduce once a year others produce several litters during a single season. Some have only one to two young at a time while others have large numbers. Most rodents are polygamous and mate for the duration of a single breeding season and some such as beavers have permanent mates. The rather high rate of breeding is intensified by the fact that in many of the smaller rodents, sexual maturity is reached at an early age, within a month of its birth normally, earlier in the females than in the males. The females breed when less than a year old. The house mouse reproduce throughout the year with an average of 5.5 litters and 31 young per female per year in building houses and 10.2 and 57 young per year on farms in the United States. Wild Norway rats are able to breed at three to four months and can produce up to seven litters a year containing 6 to 22 young. Of the species commonly found in Mizoram (*Rarrusnitider* or the Himalyan rat and *Rattus Miviventer Mentosu* being two of the main ones), the gestation period is said to be of 20 to 22 days. Considering one female rat can give a litter of 6–12 young ones and that the mother can attain immediate *post-partum* pregnancy, the rate of reproduction can be high. Further, rats can breed throughout the year and the maturity period of a young one is about 60–70 days. If one combines all this, then one family of rats—producing an average litter of eight every month and half of its litter (the female) each again producing eight from their date of maturity—can produce thousands of rats in one year. Obviously, therefore if reproduction was allowed to go unchecked and if there were no other balancing factors rats would soon rule the eco-system.[56] Therefore the reports that rodents produce prolifically after consuming the bamboo fruits were not untrue. However the answer to the riddle lies not in any one factor but in a complicated combination of factors which until discovered will remain an enigma.[57] The tribal elders from ancient times had understood the connection between the increase in rat population and the consumption of bamboo fruits and the resultant famines, which have been subsequently confirmed by the findings of modern science. But, until recently most scientists discounted even the tribal belief that famine follows the flowering of bamboo.

The Flower of Evil

The flowering of bamboo plants has traditionally been associated with impending calamities and misfortunes. In southern China, the periodic bamboo bloom was accompanied most of the time by severe floods which the people of the region felt were connected. In Abra Province in northern Philippines mass flowering of the *schizostachyum lumampao* in the late 1980s was associated with the severe earthquake that struck the area later. In August 1996 tens of thousands of floating dead rodents were reported to be floating in the lower Subansiri and Ranganadi rivers of Arunachal Pradesh in India, carrying them with the floodwater to north Lakhimpur of Assam state spreading panic and posing a serious health hazard to the people living along the course of the river.[58] An expert team, which visited the site attributed this to the flowering of *kako* species of bamboo which caused an enormous increase in the rodents in the bamboo jungles of the area. The floodwater reportedly killed and washed away portion of these population during the river course.[59] It has also been found that bamboo flowering and the phenomenon of consequent increase in rat population is perfectly known to the tribes of Arunachal Pradesh. The bamboo flowering is known in colloquial languages of the tribe as *Talamlamnam*, the increase in rat population as *buli jugnam* and the resultant famine as *duna manam*.[60] Such events in the area had taken place earlier in 1947 and before the earthquake and great flood of 1950 in Assam. Bamboo flowering is considered a bad omen not only in the Mizo Hills, but also in Assam, Arunachal Pradesh and Uttarakhand. It is often believed that great catastrophes could either precede or follow a bamboo flowering event. The 1881–2 bamboo flowering preceded the great earthquake of 1897. Similarly the earthquake of 1950 almost destroyed the entire north-east was followed by the bamboo flowering of 1955–7. While for the Mizos and Arunachali tribes, bamboo flowering is a sign of impending famine, in Assamese folk belief, the flowering of *mokal* or *kotoha* or *kotah* (a thorny species of bamboo) is a bad omen.[61] There are a number of Assamese Bihu songs which refer to the association of disasters with bamboo flowering. Similar is the belief of the Bengalis of Tripura in India, Sylhet, and Chittagong in Bangladesh which borders Mizoram and for whom this particular flowering species is known as *muli* bamboo. As early as 1836, Sir V.B. Jones connected the bamboo seeds with an impending famine. He wrote, 'For, says the Brahmins, when bamboo produce sustenance, we must look to heaven for food.'[62] *Dymock Pharma* the medical journal in their third issue in 1893 substantiated the prevalence of such beliefs in connection to the bamboo blooming. 'In India, bamboo flowering (is) an event of rare occurrence (which) has been supposed to bring in its train all sorts of evil accompanied by dire distress and famine.'[63]

The situation is being repeated prior to the impending flowering in the year 2007. Despite the modernity people still act according to these beliefs. When the Government of Mizoram declared in 2003 that 34 vegetation sites had been sighted where bamboo was found to be flowering, the tribal Christians were flocking to churches to offer special prayers to ward off famine which they feared could strike the region anytime.[64] Despite their change of faith the Mizos go by their traditional belief that when bamboo flowers, death and destruction will follow. The Chief Minister of Mizoram, Zoramthanga himself supported the act saying, 'It is not a myth or superstitious belief to think that bamboo flowering signals famine [*sic*]. It is a stark reality and we have experienced and witnessed an outbreak of famine in the past under similar circumstances.'[65] Following the flowering of these plants in parts of Tripura state contiguous to Mizoram, there has seen large-scale evacuation in 2005 as the environment created panic among the people there. It is a common belief among the poverty-stricken tribal population in the interior region of the state that bamboo flowering brings with it famine, epidemic and starvation. This fear coupled with superstition has resulted in hundreds of tribal families leaving their hamlets. The beginning of flowering has coincided with the advent of viral fever and an outbreak of gastroenteritis spreading panic. Similar panic and desertion has been reported from the Naga, Kuki and Zo tribes of

Ukhrul, Tamenglong and Churachandpur districts of Manipur state where the bamboo plants flowered with corresponding increase in rat population in mid-2004. Manipur last witnessed the phenomenon in 1954–5, around the same time as the great rat-famine in Mizoram. Confirming the fear of the people, experts stated that besides a fair chance of famine, plague and leptospirosis might also be a major threat to the people in this region.[66] According to this report what makes the situation worse is the fact that besides eating up agricultural produce the rodents also carry almost 200 types of *pathogens*, which pose a grave threat to human beings.[67] The governments in the respective areas, therefore, activated their machinery to tackle the menace.

Bamboo flowering as seen above is a recurring theme of fictions, colonial monographs and a serious topic of research for botanists. But the concurrent rat-famine has been a serious concern for the administration. Bamboo flowering indeed is an environmental activity. But the famine that results has serious social and economic repurcussions and also influenced politics in the north-eastern region. The bamboo flowering can be said to be a grand event since it ensures the survival of the species. It can be said to be celebration time for the rodents since the abundant food supply gives their population a boost. But for other species that depend on bamboos it is a calamity. The Panda which feeds on bamboo leaves has to starve or migrate, as with blossoming, bamboos begin to rot and die. The Mountain Gorillas of Rwanda, which feeds on *arundinaria alpina* bamboo leaves, also become a sorry victim of the phenomenon. It is a calamity for human beings too as most of the rural and tribal people depend on bamboo, which can be said to be the 'poor men's wood', for the construction of houses. For the tribal groups such as the Mizos, their kitchen utensils, hunting bows and arrows, spinning gadgets for textiles, pots for carrying water baskets, beds and smoking pipes too are made of bamboo. In fact, there is hardly any item, which is not made of bamboo. Hence they were hit hard. They have to construct or repair their houses, carve out their utensils and make their tools before the plants start to rot and die. The rat-famine was a huge additional hardship. Modern science has testified that the death of the bamboo plant results in soil erosion. The soil, previously bound together by the bamboo roots, erodes away affecting agricultural productivity.

Mizo oral history testified that the previous occupants of their land had migrated out of this land due to this scourge. The Mizo tribes too were forced to migrate to the hills of Manipur and Assam plain. Every famine would carry away a large number of lives with it. For a tribe poor in its demographic resources, it was a horrifying experience. Even when they survived the famine, the depopulated tribe would be vulnerable to head-hunting attacks or to subjugation and enslavement by the neighbouring tribes. But then calamities have always had such profound impact on human civilization. The number of people killed by natural calamities would be higher than the lives lost in all the battles and wars put together. Earthquakes have destroyed cities. Civilizations disappeared due to volcanic eruptions. Floods have washed away ancient cultures. Landslides have permanently buried habitats. Massive deaths and migrations resulted from famines. Diseases have eradicated habitations. Calamities might have had little space in historical research, but it has always made a tremendous impact on the making or unmaking of human history—acknowledged or un-acknowledged by historians.

The People

Keeping a wider audience in mind it is important to provide some reference to the people and region, mentioned in this book. Mizos are an Indo–Mongoloid tribal people who mostly live in the Indian province of Mizoram. The twenty-sixth state of the Indian federation, Mizoram is situated in the north-eastern corner of India. Mizoram is the land of the highlanders [Mizo: highlanders; ram: land]. It has an area of 33.9289 sq. km, of mostly hilly terrain. It is bound in the north by Cachar and Hailakandi districts of Assam, on the east and south by the Chin Hills of Myanmar (Burma) and on its western frontier lies the Indian state of Tripura and the sovereign state of Bangladesh. Mizoram is located between the 22.20′ and 24.27′ (N) latitude and 92.20′

and 94.29′ (E) longitude. It has a long international boundary touching both Myanmar and Bangladesh. The present international boundary was the creation of the colonial state during its withdrawal from the subcontinent. What is now foreign land lying across the border was hitherto the undivided ancestral lands of the scores of sub-tribes who now constitute the generic Mizo tribe. Indeed, the Partition of India in 1947 had a devastating impact on the Mizos. The Mizo tribes were distributed over a huge area in India as well as Myanmar and Bangladesh. They were a conglomeration of sovereign tribes living under the rule of their respective chiefs before the British conquest. It was British subjugation, which divided them between two diverse areas: Myanmar and India. As long as India was under British rule, their socio-economic life was not disturbed as Myanmar formed a part of the British Empire. But the separation of Myanmar from British India in 1937 and the Partition of India distributed the Mizos between three sovereign countries—India, Pakistan (East Pakistan, later Bangladesh) and Myanmar thereby rupturing the emotional, social and economic ties with each other. The boundary with Myanmar extends 434.43 km and with Bangladesh over 254.22 km. Such a long international boundary makes Mizoram strategically important. The hill ranges in Mizoram have an average height of 900 m and run in the southern direction interspersed with deep gorges in which rivers and streams flow from the high hills. There are a number of rivers, streams, brooks and waterfalls which flow to the brim in the monsoon. The most important rivers are the Tlawng (Dhaleswari) which runs 200 km, the Tuirial (Sonai) running for 150 km and Tuival running about 60 km which flow along the northern Mizoram and eventually fall into the Barak River. The southern hills drained by Chimtuipui (Kolodyne) has a course of 150 km on the east with its tributaries like Mat, Tuichang, Tiau and Tuipui while the Karnaphuli runs for 80 km with its tributaries—Tuichawng, Kap, Deh, Phairuang and Tuilianpui inundates the western Mizo Hills. The river courses are somewhat compli-cated. The Tlawng River runs for 64 km northwards while parallel to it runs the Matriver and Deh to the south. In the same way the Tuivawl, Tuichang, Tuilianpui and Tut run parallel to each other but in opposite directions.

The entire Mizoram state is constituted of hill ranges except for a small portion in the extreme north and south, which is flat. The hill ranges are generally very steep. The average height of the hills varies from 900 to 2,165 m. The Phawngpui or the Blue Mountain is the highest peak in Mizoram rising about 2,165 m situated in south Mizoram. The rest of the peaks are Lengteng (2,149 m), Naunuarzo (2,140 m), Surtlang (2,016 m), Zopuitlang (1,963 m), Lurhlang (1,935 m), Tan (1,926 m), Muifang (1,922 m), Ngurtlang (1,895 m), Tawi (1,890 m) and Rangturzo (1,855 m).

Heavy rainfall and the humid climate has turned Mizoram into a thickly forested area, but over the years the practice of *jhum* (slash and burn type of cultivation), has reduced forest cover drastically. Most commonly found trees are the different species of *chams*. The hills are covered by major varieties of bamboos (*Melanocana bambusieides*), wild banana also cover many hill slopes.

The total population of Mizoram according to the 1991 Census is 6,89,756 in which the number of males were 3,58,978 and females 3,30,778. Thus the male-female ratio is 921 females for every 1,000 males. The density of population is 32.77 per sq km. Mizoram is one of the most literate states of India. The literacy rate is 82.27 per cent according to 1991 Census, one of the highest in India. The male literacy percentage was 85.61 while the female was 78.60.

The British knew the Mizos as Lushai. It was a term used by the Bengalis in the bordering plains to describe one of the major tribes of the hills. There are a number of tribes inhabiting these hill ranges who all belong to the Kuki–Chin ethnic group. On the eve of British withdrawal these conglomeration of tribes inhabiting the hills decided to call themselves the Mizos. The word Mizo is generic term applied to a conglomeration of Chin–Lushai tribes and sub-tribes spread over Mizoram, Tripura, Manipur and sovereign states of Bangladesh and Myanmar. The names of these tribes and sub-tribes are Lushai, Pawi, Lakher, Paite, Ralte, Hmar, Vaiphei and so on. Among them the Lushais were the dominant whose dialect called *duhlian* emerged to be the *lingua franca* of all the Mizo tribes.

The origin of the word 'mizo' is shrouded in mystery. Generally these group of tribes would describe them as Zo. The word 'mizo' has been constructed recently by the Mizo intelligentsia. It has been borrowed from their own language, from the word 'mi', meaning 'people' and 'zo' meaning 'highland'; in other words 'highlander'. Lt. Col. Shakespear, who wrote the first ethnographic monograph on the tribe observed that Lushai was actually the name of single tribe—though the most powerful amongst the many tribes that inhabited the Lushai Hills. The British following the plainsmen described all the other tribes of the region as Lushai as a generic appellation which was incorrect:

> The term Lushai as we now understand it covers a great many clans: it is the result of incorrect transliterations of the word Lushai which is the name of a Clan, which under various Chiefs of the Thangur family, came into prominence in the eighteenth century. . . . In this monograph Lushai is used in the wider sense. Lushai being used for the clan of that name . . . and the general population of the hills is spoken of as Mizo.[68]

Lushai was therefore a later colonial coinage. Accordingly, the British described their habitat as Lushai Hills. Even under this name, however the tribals always called themselves Zo. It was reinvented in the 1940s to settle the identity and unity related issue of the Mizos. The word 'mizo' simply means hill-dwellers and 'Mizoram', the land of the Mizos. In 1946, during the formation of the first political party in Mizoram, the appellation Mizo was used for the first time in lieu of Lushai to cover all the tribes and sub-tribes living in the region.[69] By the end of the Second World War it was apparent that the British were going to withdraw from India in the very near future. The assumption of the Labour Party to power in Britain made it almost certain. As expected, the formal announcement of the British to leave the Indian subcontinent by June 1948 came on 20 February 1947. It prompted hectic political activity all over the subcontinent. The whirlpool of developments caught the tribals of the north-east unaware. They were generally leading a complacent life under British rule. After years of mutual warfare and bloody head-hunting raids on the plains, they had settled to a peaceful life. Their migratory movements ceased. The British kept them 'excluded' from participating in the political life of the subcontinent. Hence they not only did not participate in organized nationalist politics of India but even the nationalist leaders did not initiate them into the political process.

Now, in 1947 there was talk of British withdrawal, Indian Independence and even the British last-ditch effort to retain the control of tribal areas through the Crown Colony Plan by which the British wanted that while the rest of India became independent, the tribal areas remain under the direct administration of the British who would rule them from London. Although the tribals were resigned to the idea of 'exclusion' from all political activity, they were compelled to rethink their strategy. If the British were going to withdraw who was going to govern the tribals? What kind of Indians would fill the space vacated by the British? And then, why should India govern them? Were they Indians? These questions they confronted were political in nature and it required political processes to settle them. But the foremost issue was to settle the question of identity. Ethnically, the Mizos are closer to the people of Myanmar. In fact, half of their own tribe continued to be on the other side of the frontier. On the other hand, they had been governed as part of the British Indian Empire for at least 50 years during which they had developed an intimate association with India, which would be difficult to severe outright. They could strive to have an independent political existence. Now, to decide all these options they required a political platform too. The Lushais had organized the Young Lushai Organization (YLA) in 1930, an organization affiliated to the Christian church on a socio-religious platform. The party had to be born out of this structure though it had to be divorced from religious affiliations. YLA was formed on the line of Young Mens' Christian Association (YMCA). Organized by the Christian church of a particular denomination. But a political party had to be secular as well as the representative of people irrespective of religion, denominations sub-tribe and such affiliations. The intelligentsia also realized that the nomenclature Lushai included only one of the Zo tribes and it was politically imperative that all the tribes were unified under one political organization. Thus the name Young Lushai Association was replaced by Young Mizo Association (YMA) using Mizo as a generic term to include all the Zo tribes like Lushai, Pawi, Lakher, Paite, Ralte, Hmar, Vaiphei and others. R. Vanlawma, the first matriculate among the Mizos and the General Secretary of the YMA took the initiative. After consultation with the other Mizo intellectuals like Dahrawka, he drafted a constitution for a political party to meet the needs of the people, which were not fulfilled by the religious YMA. But a major problem that they faced was the issue of inclusion of the Mizo chiefs in the political party. The Mizo chiefs had ruled the Mizo tribes for a long time. Over the years, it had become an oppressive institution, which common Mizos wanted to be rid of. But the British perpetuated it. The inclusion of the chiefs in the party meant they would continue their tyrannical rule even after the British withdrawal. To avoid that, the party was confined to the commoners and named as Lushai Common Peoples Union with R. Vanlawma as its first general secretary. Vanlawma, on assumption, insisted on substitution of the word 'lushai' by the generic Mizo so that the party represented all the Mizo tribes not just the Lushais. Vanlawma also felt that the words 'Common Peoples' depicted the division within Mizo society, which should be done away with. Hence, it was decided that the party be renamed and formally launched as Mizo Union on 9 April 1946. This permanently settled the question of identity as well as the organization of a platform to deal with critical political concern.[70]

After Independence, on the demand of the tribe the name of the district was also changed from Lushai Hills District to Mizo Hills District by an Act of Parliament in 1954. It was then made into a union territory in January 1972. Mizoram became a full-fledged state of the Indian union on 20 February 1987.

The Event: A Prospective Famine in 2007

In 2002 there was a sudden spurt of interest in the bamboo-flowering phenomenon in Mizoram. The reason for such an all round 'interest' was the prediction by the Mizo elders that the next *mautam* was likely to strike soon. According to the Mizo predictions based on the 50-year cycle of bamboo flowering, the next famine was likely to break out in 2007. True to the predictions, as early as 2000 the 6-ft high bamboo plants started to bloom into mauve, yellow and crimson coloured flowers in the southern areas of the state conveying an ominous omen of impending doom. The alarm had started ringing among all sections of the Mizos as they were reminded of the grim days of the late 1950s when scores of people perished in hunger as the hungry rats devoured their staple crops.

The print-media was quick to pick up the issue. *The Telegraph* (Calcutta) reported, 'Mizoram Apprehends Rat-Triggered Famine' (18 March 2001). In July it had a headline 'Mizoram Devises Two-Pronged Strategy to Combat Famine' (6 July 2001). *The Shillong Times* reported 'Mizoram Government to Purchase Rat-Tails: A Unique Measure to Check Famine' (15 June 2001). *The North East Sun* had a cover story entitled 'Catastrophic Famine to Strike North-East' (1–14 July 2002). *The Eastern Panorama* (March 2002) as well as *North East Frontier Magazine* ran similar cover stories. The predicted bamboo flowering, consequent rat-famine and Mizoram as the land where this unique phenomenon would occur in 2007, brought it worldwide attention. According to *The National Geographic News* the phenomenon was intriguing as it was interesting:

> Some species of bamboo flower only every 40–50 years. In an intriguing chain of events, these periods of flowering sometimes lead to the destruction of the basic crops and widespread famine in areas of India where bamboo grows heavily. The last famine of this nature occurred in 1961 through 1965 in the hilly state of Mizoram in eastern India, an area of 21,000 square kilometres with a population of more than 7,00,000. There are hundreds of species of bamboo in the world. Some flower every year, some at regular intervals. But a small percentage flowers in synchrony, over hundreds of square kilometres every few decades. Researchers are not sure how it happens. Science has not been able to explain how the same message is passed among bamboo clumps separated by hundreds of kilometres to flower at the same time . . . [Perhaps] the rhizomes of the bamboo have some kind of memory trait that make the plants grow in synchronisation, then burst into bloom all at the same time.

After the massive flowering, the bamboo clumps die in a kind of suicide by over production . . . when bamboo plants flower, they produce a large volume of seeds, which are source of food many predators, especially rats. As masses of flowering bamboo produce this natural bounty, rats are attracted to the area. Fortified by the protein rich seeds, they multiply rapidly. But the supply of bamboo seeds is limited. When it is exhausted, armies of the marauding rodents turn their attention to standing crops, devouring acres of rice, potatoes and sweet potatoes within a few days. As a result, local peasant, who are fully dependent on agriculture for their subsistence are subjected to famine.

The phenomenon mainly occurs in Mizoram (India), which is heavily covered with two species of bamboo. The impact can be so devastating that local folklore is full of tales of this natural cyclical event.[71]

The British Broadcasting Corporation (BBC), in its online world news edition reported,

The Indian state of Mizoram is preparing itself for a mass flowering of bamboo—which has the potential to devastate the area. The bamboo only flowers once every fifty years. When it last did so, in the 1950s the abundance of seeds led to plague of rats—which in turn led to widespread famine and thousands of deaths in the region. The bamboo is produced from an underground stem. Shortly after flowering—which it does in waves—it dies. Indian bamboo expert Professor M.P. Ranjan told BBC World Service's Outlook programme that the wave of flowering is expected again in the next two years and potential disaster looms. . . . Dr. Ranjan's fellow expert Dr. Eric Bo said 'this is going to happen, it is an act of God or whatever your belief. They are going to flower and they are going to die and we know approximately when that is going to happen.[72]

The International Forest Conservation Portal recorded, 'The bamboo forests in India's north-east flower every few decades and folklore is rife with tales of devastating impact of this natural cyclical event.' It quoted an agricultural scientist who says,

Why bamboo clumps flower remains a mystery but it is believed to be caused by a characteristic that makes them grow in unison, then bloom. The flowering of bamboo produces huge seeds that attract predators, mostly rats. Strengthened by protein rich seeds the rodents multiply quickly. When the seeds are exhausted armies of rats chomp their way through rice and potato crops and granaries causing famine.[73]

The International Electronic Magazine on Environment, Energy and Sustainable Development, *Terra Green* also reported,

The north-eastern part of India is gearing up to protect itself from the possible famine triggered by a huge surge in rat population, the result of the flowering of bamboo plantations that began this month and is expected to peak in 2007. . . . Science has no explanation till date about the manner in which the message to flower at the same time is passed among bamboo clumps that are separated hundred of kilometres apart. It is speculated that the rhizomes of bamboo have some kind of 'memory' trait that makes them grow in synchronization and then burst into bloom at the same time. After massive flowering the bamboo clumps die, a sort of suicide due to over production. In previous years the simultaneous production of seeds by millions of bamboo plants caused a surge in the number of seed-eating rodents, fortified by the protein rich seeds multiplying rapidly. But the supply of bamboo seed is limited. When exhausted, armies of these marauding rodents turn their attention to nearby paddy and potato fields in search of food with devastating effects on these crops.[74]

The website of the International Human Flower Project posted a report for the surfers saying:

When Charles Baudelaire imagined Flowers of Evil, he likely never heard of Mautak bamboo. People in North eastern India's state of Mizoram refer to Melocanna baccifera as mautam or famine. The Indian peasants and officials are struggling to prevent the next flower fed disaster by harvesting the plant now and distributing poison and rattraps in the region.:The flowers produce millions of protein rich seeds that are devoured by rats causing an explosion in the rodent population. When the bamboo seed supply is exhausted, the rats move on to the next available food source—the Mizoram peasents' meagre stores of rice and cereals.[75]

A similar report prepared by Peter Foster, was published in *The Daily Telegraph* of London which described it as a 'Strange Story'.[76] Another website published a rare photograph of the bamboo fruit describing it as 'Disaster Fruit'. It reported:

Most of the bamboo species flower after the end of a long number of years of vegetative growth. The flowering is synchronous over vast tracks of landscape and this phenomenon is called gregarious flowering, which is followed by death of bamboo clumps. But such peculiar behaviour of the bamboo sometimes create problems for ecology besides human beings. The gregarious flowering of bamboo results in production of large quantities of seeds which in turn, supports population explosion in rats. However quantity of seeds available for rat diminishes soon after the germination of seeds following rains. The resultant short supply of bamboo seeds on the one hand and a large population of rats on the other lead to a phase when the rodents head towards farmlands in adjoining areas, causing widespread loss to crops. Such a chain of events has the potential to cause a famine. The last gregarious famine with a cycle of 48 years, of *muli* bamboo occurred in 1959 and the next is expected to start next year and reaching its peak in 2007.[77]

The mass of material available in the internet on the subject was indicative of the growing curiosity regarding the subject. A search of internet resources brings to notice that there is hardly any reading material from the Indian science fraternity on the issue.[78] Amusingly, and unfortunately much of the material on the phenomenon related to Mizoram, as available in the internet are unacknowledged borrowing from my paper published in the *Economic and Political Weekly* (24–31 March 2001) including its printing errors. For the Department of Science and Technology, Government of India, Centre for Science and Environment, New Delhi, and Centre for Environmental Education, the phenomenon is still 'mysterious'. Prof. B. Hari Gopal of the Department of Science and Technology, Government of India and Prof. H.Y. Mohan Rao of the University of Delhi were among the firsts to initiate scientific enquiry into it. All these indicated the governmental agencies, NGOs and botanists were equally in the dark about the phenomenon leading to a lot of speculation and unneccessary panic.

The invasion of rats and the consequent devastation from the famine was inscribed for posterity by S. Barkakaty, the first Indian Superintendent and also the Deputy Commissioner of the then Mizo Hills district of Assam (1949–53) who had occasion to experience one of the famines (1952–9):

They come in thousands and spread themselves to cover so wide an area that they eat up the entire paddy of a *jhum* overnight. When mautam year approaches, the cultivators take every possible preventive measure in their simple way by setting up traps around the *jhum* and keeping watch day and night to frighten away or kill rats. But these efforts are not sufficient for saving the crop. During such a famine (therefore) people have to depend for their sustenance and survival on wild roots, jungle fruit and anything that is edible. As a result of under nourishment various epidemics break out causing heavy loss of life.[80]

In early 2003, the Mizoram state government evolved its own contingency plan to avert a famine. It set-up a high level 'Mizoram State Rodent Control Committee' with the Chief Minister, Zoramthanga as its chairman. The seriousness with which he saw the issue can he inferred from the 2004 newspapers report. Zoramthanga, allaying the fear of the repetition of the devastation of the 1957 famine stated,

> The biggest difference between then and now is that we are no longer under Assam. We are free to follow our policies. To tackle the phenomenon we have chalked out a multi-pronged strategy that focuses on utilisation and generation of more species [of bamboo] . . . the government will not only prevent a repeat of the tragedy four decades back but actually pave the way for a bamboo revolution in the State.[81]

The committee initiated a mass campaign for killing rats in fields and houses to pre-empt their proliferation. It even offered Re. 1 for the tail of every dead rat. In the first 18 months nearly 80,000 rats were killed in the state. Two expert 'rodent killers', John Bourne and Valerie Grassmen—bio-scientists from Canada who have successfully eliminated rodents and pests in their country had also visited Mizoram in November 2002, on the invitation of the government.[82] They held extended consultations with local experts and submitted a report recommending the annihilation of rats using poison, traps and holes dug in the ground. They suggested the use of zinc phosphide and bromadiolone as a poison and said that the chemicals should be scattered in areas away from human habitation before tilling for *jhum* started.

The central government did not lag behind. It sent plant physiologist Professor H.Y. Mohan Rao of the University of Delhi to help control the situation in 2003 who recommended the farmers to plant crops that the rats did not eat such as banana, ginger and turmeric instead of the traditional items during the period when vast fields of bamboo were expected to flower. The Government of India financed a number of seminars and workshops. The recommendations of which were placed before the Planning Commission.

The Union Ministry of Environment and Forests had roped in experts including International Bamboo and Rattan Network and the United Nations Industrial Development Organization (UNIDO) for hectic consultation on how to handle the natural growth cycle of the bamboo plant which would have a direct bearing on the socio-economic well-being of the popular and a long-term ecological impact. The fact that after flowering, the entire bamboo forest die and rot was a more serious concern. Having an area of about 18,000 ha of the region covered with about 25 million tonnes of rotting bamboo was a ghastly prospect. More so, since a thriving economy revolves around bamboo. The pulp and paper industry, construction, cottage industry, handloom, food, fuel, fodder and medicine annually consume about 22 million bamboos. If left unharvested this would amount to a loss of around Rs.12,000 million. This will seriously undermine the production of paper industries of the region, as bamboo constitutes their major raw material. Bamboo rotting over hundreds of acres and growth of rat population would have a devastating effect on the *jhum* cultivation which is mainstay of the rural people of the region. Women who make the majority of the rural work-force and contribute more to holding up the rural economy would particularly be vulnerable. Their major source of cash-generating income—such as *jhum* field produce, the vegetable from the wild and the bamboo shoots which they gather and sell in town markets would disappear at least for a crucial period of time seriously affecting family budget. Again, water, which is already a scarce resource in most of the hill regions of north-east India would become scarcer. This was the experience of past years. Thus, women and children would be forced to spend longer hours to fetch water.[83] Experts testified that during the bamboo flowering in Mizoram there was a sharp increase in temperature followed by a spell of dry arid weather, which had direct fallout on the health of the people.

The impending event provided the scientific community an opportunity to study the phenomenon and evolve a strategy to pre-empt the catastrophe. The Jorhat Rain Forest Research Institute organized a workshop in April 2002 which was attended by the officials of various north-eastern states, organizations like International Network for Bamboos and Rattan (INBAR, Beizing), Gramland and Fodder Research Institute (Jhansi), Forest Survey of India (Dehradun), Institute of Wood Science and Technology (Bangalore), Himalayan Forest Research Institute, State Forest Research Institute (Itanagar), Kerala Forest Research Institute (Peechi), United Nations Industrial Development Organization (UNIDO) and so on. The workshop recommended that Mizoram should go for early harvesting of bamboos before they flowered and attracted rodents.

The idea of harvesting the entire bamboo population has appealed to the Chief Minister of Mizoram, Zoramthanga. Mizoram harvested about 40 per cent of India's 81 million tonnes of bamboo. Now with a one time harvesting the minister wanted to 'reverse the flowering phenomenon into an economic opportunity'. He felt it could bring Mizoram a 'green gold revolution through bamboo'. He proposed setting up an incense factory with machines imported from China or Thailand besides making paper and other handicraft items from bamboos. Prof. M.P. Ranjan, head of the Centre for Bamboo Initiatives at the National Institute of Design at Ahmedabad who was part of group of botanists, engineers, architects and civil servants trying to counter the catastrophe commented that it was of the 'same magnitude as a flood or an earthquake' was also equally enthusiastic about the prospect of harvesting this 'wonder grass' for an industrial purpose.

The Ministry of Environment and Forests put a dampner on the proposal stating that about 26 million tonnes of bamboo spread over more than 10 million ha was going to be affected by the imminent flowering. Only 10 per cent of these bamboos grow in accessible areas and can be retrieved for industrial use. Another major problem is the requirement of a massive storage facility to cope with the glut of such harvested bamboos. Without proper storage the bamboo is likely to quickly rot. India's Ministry of Environment and Forests set-up two committees to recommend ways to limit crop losses. One suggested that bamboo was extracted before it flowers and that mixed vegetation was planted immediately after flowering to stop soil erosion. The second recommended improving harvesting and storage facilities for extracted bamboo and removing export restrictions to find additional outlets for harvested bamboo.

In a separate proposal the Centre for Indian Bamboo Resource and Technology (CIBART) forwarded a pilot project with the Ministry of Rural Development and the Manipur state government, which involved creating a 'buffer zone' in which bamboo would be completely removed to deter rats from around bamboo growing villages. Plant Scientist Indira Khurana of the CIBART recommended that bamboo seed could be collected from the areas in which it has already flowered and immediately planted in the buffer zone. This would limit the number of seeds available to rats and would also reduce the time during which bamboo would not be available to local communities.

Some non-governmental organizations and individual environmental activists were horrified with the idea of *en masse* harvesting of bamboo plants to prevent it from flowering and resulting in famine conditions. Manorama Savur, an expert on bamboo stated,

> the first cause for alarm is not the threat of gregarious flowering which is a normal phenomenon but that a United Nations organization is ringing alarm bells while the central government is again getting caught in the UN's hysteria. The Government of India seems to have forgotten that the United Nations Food and Agriculture Organization (UNFAO) was basically responsible for both the death of the bamboo forests and the destruction of the forest ecosystem in the entire bamboo region of the southern and central India from 1960 onwards. The north-eastern region was mercifully spared, as UNFAO had not entered that region. . . . The second alarming fact is the central government's threat to harvest the bam-

boos before they flower. If an important species is clear-felled before it flowers and its rhizomes having reached the end of the physiological is exhausted, it would mean a death, a criminal murder of a species.[84]

The basic point of the activist was that flowering of bamboo is a normal reproductive activity like any other organic body. The function of a flower is to produce seed to perpetuate its species. This is true of bamboo too except that each species of the bamboo tends to flower gregariously over vast stretches of areas, but only at the end of its vegetative growth which varies from one to hundred and twenty years for different species. It implies that while any one species is flowering gregariously all other species is bamboo in that region would continue to live and vegetatively reproduce their clump. Mizoram has other species that will not flower in 2007. Thus while one species flowers and dies, others would continue to live nullifying the fear of the people about the shortage of bamboos. Again, though the fear of invasion of rodents is real there is no reason it could not be effectively tackled. Like bamboo seeds, rats also devour, wheat and rice seeds profusely. If rodents could be prevented from entering the ripe grain fields and storehouses, they could be from the ripening seeds of the bamboo too. The activist also refuted the idea that bamboo flowering portends famine. On the contrary it is the famine—which often is the result of a serial drought—that results in gregarious flowering of bamboo across the species. It is a defence mechanism—a simple survival strategy of the species to survive, reproduce and perpetuate itself. Bamboo is a grass, rigid and woody with lignin but is essentially a cellulose, carbohydrate with a large water component. Good rainfall or running water is essential for its growth. Water is both an agent of nutrient distribution in nature and a significant medium of biological acitivity in plants. Repeated failure of monsoon rings the death knell for bamboo.[85]

A retired professional forester A. Maslekar from the Indian Forest Service reported that a bamboo species (*Dendrocalamus structus*) flowered gregariously in Maharashtra in the year 1983–4 and also in Madhya Pradesh and Andhra Pradesh in the following years. He asserted that misconceptions about a natural phenomenon like bamboo flowering are perpetuated by vested interests. 'Bamboo flowering is a natural phenomenon. It is preceded by increase in rodent population—a nature's way of controlling profuse seeding. Bamboo seeds are eaten by rats, cattle as well as people. It is quite nutritious and even some believe that bamboo seed is also invigorating and has aphrodisiac qualities.'[86] He felt that under such a situation to consider the harvest of bamboo prior to flowering is neither silviculturally sound nor ecologically desirable nor is it economically welcome. It will entail wiping out of a precious natural resource altogether from the site as there will be no seeding and regeneration . . . the idea to harvest bamboos prior to flowering needs to be nipped in the bud. Everything has its time and nature has its own time-table. We have already interfered in nature's way too much.[87] The same sentiment was echoed by environmentalist P.K. Gautam 'More pressing that the US walking out of Kyoto Protocol and Russia following suit is the need to take stock of the impending anthropogenic mess that we are about to create on the subject of bamboo.'[88]

Confirming the medicinal and aphrodisiac qualities of bamboo seeds another retired civil servant from north-east J.A. Dunn, who find actually experienced the phenomenon called for scientific research on bamboo flowing:

When the (last flowering of *muli* bamboo took place, the villagers collected the seeds which are pear-shaped just like William pear—about the same size. They cooked and fried the seed and consumed with rice. After some time it was noticed that the incidence of pregnancy increased among many of the married women. Assuming that consuming the seed had somehow increased the fertility of these womenfolk, I collected a fair quantity of seeds and sent to the company along with my observations and requested them to have the seeds analysed but it appeared nothing had been done. The company however requested me to send more seeds but as it was practically the end of flowering I could collect a few kilograms only. It has been said that the rat population increased due to abundance of food, it may however be the other way round as in man. Therefore if proper research/analysis is conducted the properties of the seed would be discovered which could then be a boon for childless couples.

> The flowering of *muli* bamboo occurs once in every fifty years. One opportunity had been lost during the last flowering but it has now again presented itself and time should not be wasted for (starting) research/analysis of the seed by medical professionals and pharmaceutical firms.[89]

As can be seen the imminent bamboo flowering has activated a huge range of people, NGOs and governments. While the to-be-affected people are petrified by the prospect and are praying to survive the impending horror, the state machinery and scientists are gearing up to tackle the calamity and mitigate the imagined hardships. This is to be compared against the pre-colonial situation when the Mizo tribe tackled this calamity, which was a part of their life and history, on their own. It would be interesting to dwell on how and why they became dependent on external agencies to fight their battle or for that matter, was it the project of these agencies to create such dependence. It is intriguing because while the rest of the world is taking extreme interest in mitigating this calamity hardly any one in India is keen to gather the tribal knowledge about the phenomenon and their history of coping with it which could be recorded to provide a meaningful insight in tackling the problem. Interestingly such interest in a famine that was to come six or seven years later was truly unprecedented if one considers the usual slow or no reaction of the Indian state towards the sufferings of the people. The reasons were not far to seek. The 1959 famine in Mizoram and the apathy of the Indian state towards the misery of the Mizo people had given rise to the Mizo National Front which organized and led a 20-year long violent secessionist movement which the Government of India had a difficult time to counter. Now that the insurgency is over, the Government of India does not want to take another chance with this imminent famine. In other words, it seems that the entire 'interest' shown in pre-empting a famine was tactical, politically motivated—not really a genuine concern of a state for its citizens. If this is the politics of the post-colonial state in 2002, what was the state of politics *vis-à-vis* the famine during colonial times? How did the other NGOs that were active in the region in those days react to the calamity? There is much talk about the Mizo insurgency having been sparked off by the bamboo flowering of 1957; was it really true and how? In other words, how did the knowledge and inherent defence mechanism of a small tribal community against natural calamities change with the onset of organized state structures of the colonial state and post-colonial state? This is the subject matter of this book.

NOTES

1. For a discusion see Ranjan Chakravarty, ed., *Does Environmental History Matter? Shikar, Substence, Sustenance and the Sciences*, Kolkata: Readers Service, 2006, pp. vii–xxx. Arabindo Samanta, *Prakritik Viparjoy O Manush* (in Bangla), Calcutta: Deys' Publishing, 2003, p. 9.
2. Samanta, ibid.
3. T.S. Glickmen, D. Golding and E.D. Silverman, 'Acts of God and Acts of Man: Recent Trends in Natural Disasters and Major Industrial Accidents', Washington D.C.: Centre for Risk Management, Discussion Paper, CRM, 1992–2000; cited in Susan L. Cutter, ed., *Environmental Risk and Hazards*, New Jersey, 1994.
4. United Nations Environment Programme, *Environmental Data Report*, Oxford, 1991.
5. In India a number of calamities are not listed as a natural calamity, e.g. bamboo flowering related famine which affect a marginal people like the Mizo tribal. Similarly not all calamities are national in character. While earthquakes in Gujarat were declared to be a calamity of national dimension, perpetual floods in Bihar, Assam and West Bengal were never considered so despite the demand of the political elite from the region. This happened because the political elite from this region were not powerful enough to make themselves heard.
6. Samanta, op. cit.
7. Susan L. Cutter, *Living with Risks: The Geography of Technological Hazards*, London, 1993; cited in Samanta, op.cit.
8. I.A. Rezanov, *Catastrophe in Earth's History,* Moscow: Progress Publishers., 1980, p. 7.

9. Rajen Saikia, *Social and Economic History of Assam 1853–1921*, Delhi: Manohar, 2001, pp. 99–104.
10. John McCosh, *Topography of Assam*, London, 1837, rpt., Spectrum, Delhi, 1986, p. 52.
11. The cause of food shortage was draught and consequent under-production.
12. Y.S. Singh, 'Nupilan: Manipur Womens' Agitation', in *Economic and Political Weekly*, vol. II, no. 8, 21 February 1978.
13. Sajal Nag, *Contesting Marginality: Ethnicity, Insurgency and Subnationalism*, Delhi: Manohar, 2002, pp. 99–100.
14. Sajal Nag, 'Rats, Tribal, State and Nation', in *Economic and Political Weekly*, vol. 34, no. 5, 24–31 March 2001.
15. J.D. Baveja, *Mizoram: The Land Where Bamboo Flowers*, Gauhati: Assam Publication Board, 1970.
16. Vasanthi, *When Bamboo Blossoms*, English tr. Gomathi Narayanam, Gauhati: Spectrum, 1989.
17. D. Rokhuma is the Grand Old Man in this state and has been the recipient of the high civilian award of Padmashree by the Indian government. He advises students, researchers, academics, policy-makers and the administration alike on the issue of bamboo flowering. He is an authority on the subject though he was a school dropout. He has seen quite a few occurrences of this reproductive cycle of the bamboo as well as the devastation it brings to the vegetation of the area and the misery to the popular. He has been a keen observer and with a curious mind. He has been able to test the peoples beliefs associated with the phenomenon against real happenings. It has tried to understand the organic correlation, if any, between bamboo blossoms and growth of rat population. As a result of his intrinsic interest in the subject, he has developed a laboratory in the backyard of his massive house where he has preserved a rare bamboo fruit, massive numbers of rat-tails and associated animate and inanimate objects as evidence from past events of bamboo flowering. Not only this, he has been a veteran crusader against the famine that results from this environmental event. He had organized a NGO called Anti-Famine Campaign Organization to raise peoples' awareness about the phenomenon and initiate action long ahead of the predicted event. He has rallied youth groups, students, ex-army men and bureaucrats around the organization. No wonder, the state administration relies more

on him to tackle the problem than its own pool of scientists and policy makers.

18. International Network for Bamboo and Rattan, http:/www.inbar.int/flowering/Assets/Quiz%20answers,htm
19. Ibid.
20. K.K. Seethalakshmi and M.S. Mukesh Kumar, *Bamboos of India: A Compendium*, Thiruvananthapuram: Kerala Forest Research Institute, 1982, pp. 1–8.
21. Ibid.
22. Shiv Vishwanathan, 'House of Bamboo', in *The Carnival of Science,* Delhi: Oxford University Press, 1997, pp. 204–5.
23. Ibid.
24. E.C.S. Baker, 'The Game Birds of India Burma and Ceylon', in *Journal of Bombay Natural History Society,* 24, 1916, pp. 201–3; cited in Vishwanathan, op. cit.
25. H.R. Blanford, 'Note on Operations in Bamboo Flowered Areas in Kartha Division', *Indian Forester,* 44, 1918, pp. 550–60; cited in Vishwanathan, op. cit.
26. E. Blatter, 'The Flowering of the Bamboo (part I)', *Journal of Bombay Natural History Society*, 33, 1929, pp. 899–992; cited in Vishwanathan, op. cit.
27. J.W. Bradley, 'Flowering of Kija Thanung Bamboo (Bambusa Polymorpha in Ironme Division, Burma', *Indian Forester,* 25, 1899, pp. 1–25; cited in Vishwanathan, op. cit.
28. D. Brandis, 'Biological Notes on Indian Bamboos', *Indian Forester,* 25, 1899, pp. 25–50; cited in Vishwanathan, op. cit.
29. T.F. Bourdillon, 'Seeding of the Bamboo Theory', *Indian Forester,* 21, 1895, pp. 228-9; cited in Vishwanathan, op. cit.
30. H. Santapu, 'The Flowering of Strobilantnes', in *Journal of Bombay Natural History Society,* 44, 1944, pp. 605–6; cited in Vishwanathan, op. cit.
31. D. Chatterjee, 'Bamboo Fruits', *Journal of Bombay Natural History Society,* 57, 1960, pp. 451–3; cited in Vishwanathan, op. cit.
32. International Network for Bamboo and Rattan, http:/www.inbar.int/flowering/Assets/Quiz%20answers,htm
33. D. Janzen, 'Why do Bamboos Wait so Long to Flower?' *Annual Review of Ecology and Systematics,* 7, 1976, pp. 347–91; cited in Vishwanathan, op. cit.
34. Vishwanathan, op. cit., p. 206.

35. *The World of Bamboo*, Research and Development Center of Bamboo, downloaded from its website, www7.ocn.ne.jp
36. As in note 18.
37. Ibid.
38. Ibid.
39. Ibid.
40. As in note 20.
41. Anil Agarwal, 'The Bamboo Famine in Mizoram', in *The State of India's Environment: A Citizen's Report*, Delhi: Centre for Science and Environment, 1982, p. 41.
42. Ibid.
43. C.E. Low, *Gazetteer of Balaghat District*, cited in F.C. Hennikker, *Maotam in Lusai: Notes Compiled in July 1912*, Maidstone, 1930, p. 6. Hennikker Papers, Box no. 10, Cambridge: Centre for South Asian Studies, University of Cambridge, p. 1.
44. Agarwal, op. cit.
45. 'Famines', in *International Encyclopaedia of Social Sciences,* ed. David L. Sills, vol. 5, New York: Macmillan and Free Press, 1968, pp. 322–7.
46. Ibid.
47. *The New Encyclopaedia Britannica*, vol. 23, Chicago: Encyclopaedia Britannica Inc, 1994, pp. 401–12.
48. Vishwanathan, op. cit., p. 207.
49. Agarwal, op. cit.
50. As in note 47, vol. 1, p. 744.
51. Ibid.
52. C. Rokhuma, *What is Anti-Famine Organization Doing*? Aizawl: Author, pp. 131–2.
53. Reported in *The Telegraph*, Calcutta, 6 July 2001.
54. Agarwal, op. cit.
55. As in note 47.
56. Agarwal, op. cit.
57. Ibid.
58. *The Assam Tribune*, 20 August 1991. *Pratidin* (in Assamese), 17 August 1996. *The Sentinel*, 15 August 1996.
59. Ibid.
60. Interview with Dr. Tai Niyori, reported by S.K. Barpujari, 'Bamboo Flowering in Mizoram: A Historical Review', paper presented in *North East India History Association Annual Conference*, Aizawl, 1996.

61. Ibid.
62. *Journal of Agricultural and Horticultural Society of India*, XIII, pt. I, Proceeding IV, 1836, cited in *North East Sun*, vol. 7, no. 23, 1–14 July 2002, p. 13.
63. Ibid.
64. 'Bamboo Flowers: Mizoram Prays to Ward off Famine', *Shillong Times,* 17 March 2004.
65. Ibid.
66. M. Bhattacharjee, 'Famine to Hit North East Soon', *Asian Age*, 20 September 2003.
67. Ibid.
68. J. Shakespear, *The Lushai–Kuki Clans*, London: Macmillan, 1921; rpt., Kolkata: Firma KLM, 2002, Introduction, paras 3–4.
69. For the details on the process of construction of Mizo identity, see Sajal Nag, *India and North East India: Mind, Politcs and the Process of Integration 1946-1950*, Delhi: Regency, 1998, pp. 25–32.
70. Ibid.
71. India Guards for Famine linked with Flowering of Bamboo by Pallav Bagla, *National Geographic News*, 22 June 2001.
72. 'Bamboo Puts India on Famine Alert', BBC News World Edition, http://news.bbc.co.uk/2/hi/south_asia/3680714.stm
73. http://www.forests.org
74. http://www.teriin.org/terragreen
75. http://hfp.srv3.pmachinehosting.com/index.php/weblog/comments/indias_flower_of_pestilence/
76. Report by Peter Foster, *The Daily Telegraph,* London, 15 October 2004.
77. Sujit Chakravarty, 'Disaster Fruits', in http://www.uniindia.com/unlive/unisite.nsf
78. The exception being H. Lalramnghinglova, 'The Coincidence of Bamboo Flowering and Famine in Mizoram', in P. Shanmughavel, R.S. Peddappaiah and W. Liese, eds., *Recent Advances in Bamboo Research,* Jodhpur: Scientific, 2003.
79. Agarwal, op. cit., also 'Bamboo Flowering and Rats in Mizoram', *Science Reporter*, July 1979, pp. 484–6.
80. S. Barkakaty, *Tribes of Assam*, Delhi: National Book Trust, 1969, p. 95.
81. Rajiv Bhattacharya, 'Mizoram on Famine Alert', *The Telegraph*, 9 November 2004.
82. Santanu Ghosh, 'Pied Pipers Come to Mizoram's Rescue', *The Telegraph,* North-East page, 15 May 2003.

83. Linda Chhakachhuyak, 'When Bamboo Flowers', in www: IndiaNest.com
84. Manorama Savur, 'Saving the Eco-System of North East', in *Economic and Political Weekly*, 15–22 November 2003, p. 4916.
85. Ibid.
86. A.R. Maslekar, 'More on Bamboo Flowering', *Economic and Political Weekly*, 13–20 December 2003, pp. 5218–5304.
87. Ibid.
88. P.K. Gautam, 'Don't Nip at the Bud', *Down to Earth*, 16 February 2004.
89. J.A. Dunn, 'Bamboo Seed and Fertility', Letters to the Editor, *Shillong Times*, 12 June 2004.

CHAPTER 2

Famine as a Site for Production of Knowledge

It was 16 April 1844. As the day ended and darkness intensified, the sleeping villagers of the Kachubari village in the Sylhet district of Bengal, on the foothills of the Mizo Hills (then known as Lushai Hills) woke up to the shrieks and attacks of the marauding tribal raiders from the hills. It was the head-hunters from the Mizo Hills who were on one of their head-hunting raids on the plains. Although the raid was not unfamiliar to the villagers, they were too surprised to attempt escape, leave alone offer resistance. In a sustained orgy of merciless violence for the whole night the raiders destroyed the village altogether. They killed a large number of villagers and carried at least 20 heads with them. Six persons including a minor girl were captured alive as captives.[1]

The news of this massacre perpetrated on a people who were now British subjects reached the authorities at the district headquarters. The District Magistrate, Sealy rushed to the spot to inquire into the episode. The investigation revealed that it was a head-hunting raid committed by a neighbouring tribe called the Paite, living in the adjacent hills. It also transpired that the ruling, powerful Paite chief Lalina had just expired and according to the burial rites of the Paites, the tribe required a number of human heads to be buried with the late chief. The raid on Kachubari village was meant to procure the heads. The heads carried by the raiders would be used for the purpose and the captives would be declared slaves. The British also came to know that such raids by these head-hunting tribes were frequent in the area where the hills merge with the plains. This attack was not only the first of its kind since the British takeover of the administration of Bengal, it was also the first encounter of the British with any real head-hunting tribes in India. This encounter proved decisive as far as the Mizos were concerned as the Paite—which perpetrated the raid, was only a sub-tribe of the generic Mizo tribe. Raids, plunder, kidnapping and head-hunting in the foothills were a part of tribal life. The British were prompt to label such acts as barbaric, inhumane and primitive. For the tribes, however, this was their way of living.

Historical compulsions like physical and numerical weakness, vulnerability to hostile attacks from stronger neighbours and search for safer settlements drove these migratory tribes to the forbidding hills, but they remained dependent on the plains. They would come down to the foothills to barter their forests, agricultural and handicraft products for salt, iron and other such items scarce in the hills. Ironically the tribes also raided and perpetrated violence on the same people with whom they generally had an amicable exchange relationship. The raids were essential for them. Raids were committed to procure items, which they could not afford to barter or otherwise procure. It was also a show of might and a sort of collection of taxes from the area. Raids were not perpetrated on the plainsmen alone. In fact, the neighbouring villages and tribes were subjected to more frequent raids. The reasons for such raids were stated to be, a private quarrel with a neighbouring clan, a scarcity of women and domestic servants and the consequent necessity of procuring a requisite number of captives to supply the wants of the tribe, the simple desire of plunder or of obtaining heads to grace the obsequies of some departed chieftain were the principal causes of these raids.[2] Kidnappings were orchestrated to procure slaves[3] who would work for the manpower—short tribal economy. It was also a way of procuring technology. The Mizos have themselves recorded that during their raids they looked for steel, which could be moulded and shaped into weapons as well as artisan and craftsmen from the plains.[4] Kidnapping was also necessary to procure women and concubines for the chiefs who needed to demonstrate their might and status by such acts.[5]

Similarly head-hunting was necessary to procure human heads

for the mortuarial rites of the chiefs or evidence of bravery and strength of a contending chief. The old monarchies—either Bengal, Tripura or Cachar—under whose suzerainty these foothill areas fell also never intervened or came to the rescue of their frontier subjects. The raiding tribes therefore never had to confront any counter-retaliation to their acts. But times had changed without their knowledge. The British had replaced the old monarchical or nazimate regimes. These new rulers were not only the avowed apostles of 'modernity' and 'civilization' but also self-righteous. They formulated their own principles and parameters of administration and ethics. Unlike the previous regimes they claimed responsibility for the defence of their frontiers and its subjects. Therefore as soon as the reality of Kachubari massacre unfolded, they were prompt to banish the acts of the raiders as barbarian and savage and set out to punish its perpetrators. The first task was to find out whether these tribes were under the political authority of any of the regional states so that the responsibility of guilt could be fixed. When it was confirmed that the tribes were more or less independent, the Government of India sanctioned the use of the military to punish the marauders, an expedition to be headed by Captain Blackwood, of the Sylhet Light Infantry. This was the first of the series of expeditions that were launched by the British, which continued up to the end of the nineteenth century. The first expedition was meant to punish the perpetrators of a gruesome murder. But the punitive expeditions subsequently transformed into wars of conquest as the British realized that the tribal raids would not cease unless they were totally conquered and placed under the 'civilized' administration of the British and culturally transformed through the quietening influence of Christianity. Hence the initial objective of 'punishment' was subsequently and silently shifted to 'conquest'. But neither punishment nor conquest of the Mizos were easy even for the mighty British with their massive manpower and modern war technology. The British tried all methods: outright war and attack but failed as the tribes resorted to guerrilla tactics; they tried economic blockade but the tribes would promptly shift to another part of the hill range. The British would destroy village after village, burn the stored food and standing corps and take prisoners as security against the submission of tribes. But the tribes did the same to the British in another area.[6] This explains why the British war against the Mizos continued from 1832 to the end of the century. In fact, no other British war (except with the Nagas) in India continued for so long. In fact, even when in 1898 the Lushai Hills were more or less completely conquered and colonial administration had been inaugurated in Aizawl village of the Lushai Hills, the Mizos continued to resist the colonial penetration in other parts of the hills. But a closer study shows that the collective Mizo resistance had begun to crumble from the early 1880s itself. The British military officers did not fail to notice it.[7] The Mizo hostility to the British dwindled. Many of the chiefs desired truce at least temporarily; warriors surrendered arms and were begging for food. The underground Mizos were coming out and seeking refuge in British territory. This was an astonishing metamorphosis of an enemy; though the development was extremely welcome the British were bewildered as to what had caused the turnround. It was not before long that they understood the reason. In fact, they saw it with their own eyes; it was the great famine or *tampuii mithi* as the Mizos called it.

Exactly 30 years later another famine struck the hills in 1911–12 resulting in similar hardships. The only difference this time was that the Mizos were under a 'foreign paternalistic' government who extended succour to the famine affected people of the hills.

Early Famines

The British came into contact with the head-hunting Kuki–Lushai tribes with the acquisition of Chittagong in 1760 from the Nawab of Bengal. This contact turned into confrontation in 1832 when Cachar was annexed to the empire. After a series of violent and bloody expeditions and warfare that lasted more than half a century, the British entered the Mizo Hills and set-up a rudimentary colonial administration to rule the subjugated tribes. It was the British colonizers who were the first to record the bamboo flowering related famine in 1860–1 when 'many died due to starvation and those who were bulky enough to survive became lean and thin beyond description'.[8] This was corroborated by a contemporary British report from neighbouring Manipur,[9] which was equally affected by it as well as the *Note on Maotam* prepared by F.C. Hennikker.[10] Although a major famine occurred in 1861–2, even though they were in contact with the Mizo tribes, the British did not give much importance to it. The reason perhaps was that the phenomenon did not affect them the way latter ones did. The British records are quite silent about it except for a peripheral mention, but the oral testimony as well as the periodicity of bamboo flowering suggests there was a severe *mautam* (famine) in 1861–2. The bamboo flowering phenomenon of 1861 in the Mizo Hills had also affected southern Manipur and the Cachar region of southern Assam, which are contiguous to Mizo Hills. Colonel McCulloch reported that by 1858 bamboo plants of the region had already flowered and the consequent increase in rodent population had caused panic amongst the Kabui Naga inhabitants of the area.[11] The oral testimonials documented that there were substantial mortality due to the famine. It also proved that the coping mechanism of the Mizos against the recurring famines was quite effective and even though 1861–2 were crucial years in their struggle against the British, they not only continued the resistance but at the same time fought and survived the natural calamity as well. It also demonstrated the severity of the subsequent 1881 famine, which not only crumbled their resistance but also compelled them to take shelter with their enemies. The British recorded the famine of 1881–2 in minute detail. It was of special significance to the British, since the famine broke the back of the Mizo resistance and brought them within the ambit of British. It was a widespread famine in which an estimated 15,000 Mizos perished. In fact, it is said that the Mizos had resisted the British for about 40 years. The 1881 famine had devastated and debilitated them so much that they easily surrendered to colonial subjugation.

There were others too like the Christian missionaries and European tea planters who also lent a helping hand to the suffering tribals. The pattern followed in yet another famine that hit the

Mizo Hills in 1930–1. Thus three successive famines resulting from bamboo flowering hit the tribals during 1881 to 1931. These famines not only proved to be a site of production of knowledge about an intriguing ecological phenomenon but also a site of a range of political interventions.

BAMBOO FLOWERING AND FAMINE: PRODUCTION OF KNOWLEDGE

Famines were not new to the British in India. They had seen quite a few devastating famines since they took over Bengal. In fact, they were responsible for most of them. They were also familiar with famine, as European history is replete with instances. But what they observed in the Mizo Hills was unique; it was an amazing ecological phenomenon. For the British it was a discovery of a new kind of famine, which stalked the hills of southern Assam comprising the Mizo Hills. The entire hill range covering Chittagong Hills, Lushai Hills and Chin Hills and Burma were covered with thick vegetation of various species of bamboo plants. The bamboo plants have a peculiar reproductive process. All species of plants flower at about the same time at lengthy intervals. This flowering cycle can be about 30, 50 or even 120 years for certain species. The flower gives way to a pear-shaped green coloured fruit which contained oval shaped white colour seed. Once the seeds sprouted, the plants began to die. A glut of bamboo fruits incites an explosion in the population of rodents that eat the fruit. The tribals believed that it was the fruit that cause increased fertility in the rodent population, which survive on bamboo fruits. But when the fruit sprouts, food becomes scarce for the rat population. Desperate for food they attack the stored food of the tribes and even the standing crops. When they finish with the human food they die of hunger but the shortage they cause to human food results in famine and starvation deaths to human beings as well. Such famines were so widespread and acute that it would take a great toll of human lives and the Mizos generally dreaded it. The two varieties of bamboo species, which are called *thing* and *mau* by the Mizos, flower in cycles of 30 and 50 years.

The 30-yearly cycles called *thingtam*, a dreaded famine for the Mizos occurred in 1881–2. It forced them to give up the 40 years resistance to the colonizers and also surrender their pride and arms for food.

The tribals produced the entire information about the ecological phenomenon of Bamboo flowering and the consequent rat-famine themselves. The Mizo Hills were steep mountain ranges inhabited by head-hunting tribes. It existed as a land, to use the words of a missionary, 'lost to civilization'. Though the tribals often came down to the foothills, plainsmen were never allowed to penetrate the thick jungle except the few tradesmen called *karbari*. As a result there were no information or knowledge about the people or their surroundings to plainsmen except the reports by these *karbaris*, which were often exaggerated. These *karbaris* picked up the language of the Mizos and were allowed access to the tribal world deep in the hills. The British recruited the *karbaris* as mediators and interlocutors during their long encounter with the Mizos. Rai Bahadur Hari Charan Sharma was one such interlocutor a *muktear* by profession who had intimate knowledge about the Mizos, utilized by the British in their often turbulent relationship with the hill tribe. They promoted him as the Special Extra Assistant Commissioner of Cachar district and rewarded him with the title of Rai Bahadur for his services. The year 1881 was a significant year in the Anglo–Mizo relationship. There was intensification of the hostile attacks from the Mizos in the early part of 1881–2 in which more than 500 people were either killed or taken captive. But by October there were signs of distress within the hills and soon the raiders appeared in British territory as refugees seeking food and shelter. The sudden turn of events bewildered the new colonizers. They ascertained from the refugees that the reason for their migration was a severe widespread famine that had struck the hills. The British deputed their trusted officer Rai Bahadur Hari Charan for verification. Once it was established that the tribals were stating facts the local authorities compiled its report, which was the first recording of the event of bamboo flowering and the consequent famine in Mizo Hills. Although there are records of bamboo flowering by botanists like J.W. Bradley[12] in other regions, this report was the first to record the entire phenomenon in its sequential development in north-east. In fact, even Bradley's report on the same phenomenon came later in 1899.

From the experience of repeated famines the Mizos also learnt to differentiate two different varieties of famines—*thingtam* and *mautam*. The former occurred due to the flowering of one variety of bamboo plants which the Mizos called *thing* and other *mau*. The former occurred in about a 30-year cycle and was less severe in its impact than the other. Moreover *thingtam* was more localized in its impact occurring mainly in the areas west of Langkaih River from where it generally spread to the rest of the country over a course of a year. *Mautam* on the other hand affected the areas east of Tuirini River before actually hitting the rest of the region.[13] *Mautam* attracted a larger number of rats and lasted longer while *thingtam* was of shorter duration. It was also noticed that the *thingtam* famine did not affect the Lakher tribes of hills. The reason was that the *thing* variety of bamboo was not found in those areas.

The tribal elders also confirmed that the arrival of *mautam* was always preceded by the swarming of an insect called *thangang* (a species of locusts).[14] The Mizos took the arrival of locusts as a sure sign of an imminent famine. The swarming generally starts at dusk and continues till late night, slowly moving towards the high mountains. The swarms would create a strong buzzing sound similar to the sounds of a monsoon cloud. Eyewitnesses recorded that the insect was the size of a grain of a corn, dark brown in colour and spotted. The trees of the entire hills would be full of these insects. Zoologists believe that the flowering of the bamboo plants led to the migration of these insects into the region. Generally the years of such famines were also a year of extreme climatic conditions like heavy rainfall or drought, which damaged the crop of the year. The famine also brings other hardships as it was inevitably followed by dysentery and cholera causing further havoc of the people. Though the people believed it was part of the calamity it was actually because in the absence of proper human food the people would eat just about anything to satisfy their hunger during the famines. The havoc caused by each of these famines were therefore a milestone in the oral history of the tribals and the survivors of these calamities use them to mark dates in their life.

Codification of Knowledge: Bamboo Flowering and Rat-Famine

The first report on the episode of the bamboo flowering and famine was found in the reports of Captain McCuloch from Manipur. It was the 1861 famine, which had spread to the contiguous areas of Manipur. But commenting on the report, F.C. Henniker, the Superintendent of Lushai Hills, 1912–13 wrote, 'I gather that Colonel McCuloch had not actually seen the phenomenon of which he writes. He was describing it from native accounts.' It was followed by the *General Administration Report of the Cachar District of 1861–2*, which also recorded the event in the Lushai Hills. Hennikker commented on this report: 'The author of the Cachar reports gives interesting information but did not discover all events and failed to record for our information what kind of bamboo flowered. It is evident that all bamboos had not seeded and died for he proceeds to speak of a large export of bamboos which was no doubt the *mau* or *muli*, a staple export in normal years from this part of the country.'[15] Subsequent to this a reference of the phenomenon of bamboo flowering was found in Sir D. Brandis' *Indian Trees* wherein it was recorded that 'there are indications that in dry stony places and in exceptionally dry seasons bamboo flower earlier and more abundantly'.[16] In the Balaghat district of Central Provinces also the blossoming of bamboo plants was recorded in 1869. C.E. Low writes in the *Gazeteer of Balaghat District* (Central Province): 'In 1869 every *kattang* tree (*Bambusa arundinacae*) in the district seeded and the green of their foliage was changed to a silvery white; hundreds of tons worth of seed were collected and eaten by the people who were then in the throes of a severe famine.'[17] It proved that not only were there bamboo flowering in other parts of India but, like the Mizo Hills there was famine in those regions too.

But in no reports any suggestions were made about the interrelation between the two occurrences.[18] The case of the Lushai Hills were not mentioned in any of the scientific literature of this period because it was not yet a part of the British Empire in India.

The first colonial attempt to codify the knowledge of famine related to bamboo flowing was done by J. Knox Wright, the Deputy Commissioner of Cachar district. During the famine of 1881, be prepared a report on the commotion in Cachar district, which had resulted from the migration of the dreaded headhunters in the valley in search of food. The report was based on his interview with the Lushai chiefs who had come to him for assistance and the diary of his deputy Hari Charan Sharma, who was sent to ascertain the events in the hills. This report was to became the basis of understanding the phenomenon of bamboo flowering and its connection with rat-famine. Knox Wright in his report stated,

> After certain intervals of time the bamboo plants swell considerably and a sort of seed is formed within them resembling ordinary paddy seed. The existence of such rich supply attracts rats in swarms and as these animals were naturally very prolific, abundance of food causes a still greater number to appear, which increase in a geometrical proportion. The rats then spread and consume everything that is eatable, sparing neither paddy crops, *kachoo* (arum), nor even cottonseeds. . . .[19]

It is evident that Wright had not yet fully understood the phenomenon. He had gathered his knowledge from the oral testimony of the tribal chiefs through an interlocutor who probably had not himself comprehended it properly or failed to translate correctly in English to the Deputy Commissioner. For nowhere in the report Knox Wright mentions flowering of the Bamboo nor does he report its fruit. He merely talked of the 'swelling' of bamboo plants, which was not the correct situation. His idea on the size of the seeds was also erroneous. The tribal chiefs also could not convince him of the fertility quality of the bamboo seeds, which caused the proliferation of rat population. It was also possible that the rational mind of the officer was not convinced of the connection between bamboo seeds and increase in rat population. Therefore he took recourse to a more scientific explanation that it was the increase in food supply that caused the increase in jungle rats.

The Annual General Administration Report of the Government of Assam also prepared an account which became the basis of the document prepared by Captain H. Brown entitled *The Lushais 1878–89* and Alexander Mackenzie entitled *History of the Relations of the Government with the Hill Tribes of North-Eastern Frontier of Bengal* in 1884. Henry Brown was the first Superintendent of the Lushai Hills (May to September 1890) who was killed by the Mizos near Changsil in the autumn of 1890. Similarly, Alexander Mackenzie was in-charge of political correspondence of the Bengal government during the period 1866–73. During this period, on the request of the Government of Bengal he prepared *for office purposes* a Memorandum on the north-east frontier of Bengal. This was later updated to constitute the above book, which again was 'meant to be useful to government and its officers, nothing more'.[20] The original official report stated, 'The famine arose according to the testimony of all persons concerned from the depredation of rats. In the pervious season the bamboos had seeded, and supply of food thus provided caused an immense multiplication in the number of rats, who when they had exhausted the bamboo seed fell upon the rice crops and devoured them.'[21] Alexander Mackenzie did not have any personal encounter with either the phenomenon or the people affected by it. He entirely relied on the official papers and the report prepared by the government. Hence, his reports too did not refer to the flowering phenomenon of bamboo plants or the rat–bamboo connection. Since then, this report was used by most of the British officers while reporting or dealing with the problem as can be seen in the military report on the 'Chin-Lushai Country' by Colonel E.B. Elly. In fact, Elly's report was a virtual reproduction of Alexander Mackenzie's, which in turn was a reproduction of the *Annual Administrative Report*, prepared by the government.

The missionaries created a more concrete knowledge base on this ecological event during the next famine of 1911–12. Sponsored by the Arthington Aborigine Mission, two missionaries J.H. Lorraine and E.W. Savidge landed in Mizo Hill on 11 January

1894. Both the missionaries were quick to adjust to the realities of the area and picked up the dominant language as well. In fact, they were the joint authors of the *First Grammar and Dictionary of the Lushai Language*. The two missionaries missed the famine of 1861 and 1881. They however, witnessed the bamboo flowering event that occurred in 1911–12. Citing the report of Alexander Mackenzie about the event of 1881, Lorraine wrote,

> Exactly the same thing happened before and after the seeding of the *mau* bamboos in 1911 and it is because I lived through the heartbreaking experiences of the *mautam* famine of 1911–12 (when the late lamented Colonel Kennedy—then Superintendent, I was privileged to assist the authorities in alleviating the sufferings of the people in some measure) that I am so desirous that the government should do something to provide against the *thingtam* famine which according to universal Lushai belief is fast approaching.[22]

On the basis knowledge procured from the Lushai elders as well as missionaries like J.H. Lorraine, F.C. Henniker[23] the Superintendent of Lushai Hills, 1912–13 painstakingly prepared the first authentic account of the bamboo flowering phenomenon. In fact, both Hennikker and the missionaries were amazed that the Mizo elders could correctly identify each of the bamboo species.[24] Hennikker was the Superintendent of Lushai Hill during the *thingtam* famine of 1911–12. He was not only stationed in the Mizo Hills to actually witness it, but was also familiar with the literature produced on it. In his note, Hennikker also prepared a list of the species of bamboo plants with their botanical names that grew in Mizo Hills along with their properties:

BAMBOOS OF LUSHAI HILLS

Mizo Name	Botanical details
1. *Mau, tak*	*Melocanno bambusoides* (*muli*) The commonest bamboo of the hills. Grows as if sown broadcast, i.e. not in clumps. Flowered seed and died in 1911 (began 1910). New crop was spouting in 1912. In 1911 there was a complete failure

	of the crops owing to rats which first fed on the bamboo fruit, multiplied and then attacked the growing paddy. The bamboo fruit is shaped like a fig or pear but the stalk is at the thick end. (The scarcity engendered by the failure of the crops is called *mautam*.)
2. *Phulrua*	Flowered and seeded concurrently with the *mao* and died. The seed is edible and said to be as good eating as maize. The seed has an envelope or husk. The seed shines and pearl-like. A big bamboo with stout stem. The biggest bamboo fall. Tufted.
3. *Rawthing*	Blue-ish under side of leaf- also stem. (When this flowers and seeds there is *thingtam*. Last Occurred 1885.)
4. *Rawnal*	Leaves same colour on both sides. Joints flush with stem—not projecting. Grows at Low altitude. Saw a clump near Shaja.
5. *Rawtha*	Tall and graceful. Tapers very gradually. Long joints. Tall. Not very thick.
6. *Rawmi*	....
7. *Turshingh*	....
8. *Talan*	....
9. *Ankuar*	....
10. *Vai-rua*	....
11. *Rawngal*	Tufted leaf. Broad and coarse and ribbed. Very dark or very dark green. Grows in forest. Droops and leans for support on trees. Resembles Savil but bigger. Seen in Karimganj at 4,000–5,000 ft. Very brittle. Would not make ropes or bow-strings. Said to have seeded since British occupation, i.e. 1890.
12. *Rawte*	....
13. *Chal*	A dwarf. Grows at a high altitude. Single.
14. *Phar*	Single. A Dwarf bamboo. Small narrow leaves. Short joints with a collar of thorns at each notch. Found on Huinjag 4,800 ft.
15. *Likh*	....
16. *Nat*	....
17. *Sairil*	A climbing bamboo. Used for bowstrings. Seen at 3,500 ft.

Hennikker related the ecological event of 1911 saying that two of the many varieties of bamboos—*mautak* and *phulrua*—flowered and seeded in 1911 and after flowering the parent plant died. Borrowing from Lorrain's *Lushai Dictionary* he took *mautam* to mean 'The Finish of Mao Bamboo'. He added that the term connoted scarcity which invariably ensued and in fact was more often used to denote scarcity than the flowering of bamboo. Tracing government records as well as the past experiences of government servants he was convinced of the connection between flowering of bamboo and the subsequent famine. But when enquired of its scientific basis, F.J.F. Shaw, the Imperial mycologist at the Pusa Agricultural Research Institute rejected the idea.[25] He explained that drought which was usually a forerunner and cause of famine, also cause or tended to cause the bamboo to flower. Hennikker asserted that there was no drought in Mizo Hills preceding 1911 either to cause the *mau* to flower or facilitate the breeding of rats. In fact, there was good rainfall in that year.

Hennikker then discussed some of the bamboo fruits, which could be eaten by human beings, e.g. the seed of the *phulrua,* which was different in appearance from the seed of *mau* bamboo. The food of the rats, however, was the pear-shaped fruit of the *mau* bamboo. From personal experience he narrated that the bamboo seed was a food eagerly eaten by rats, which thereupon multiply enormously and it was these rats that devastated the crops of 1911. It also became evident that *mautam* was 'not over in a day or even a year'. For example, the bamboo began to flower in 1910 and the famine continued from 1911 to 1912. He also corroborated the Mizo belief that the swarming of the *thangang* insect signified imminent famine. But Hennikker felt that it was these insects, which were responsible for the damage of the crop along with the rats. *Thangang* was however equalized with 'a sort of cockroach' by him instead of the brown locusts as was traditionally believed by others. He felt that as per the periodicity of the bamboo flowering and its reproductive cycle the next such event was likely to strike the Mizo Hills around 1930-1. He also lamented that the official records for the phenomenon was defective and the researches of the botanist could not be relied on as it showed indefinite results. As a result the exact dates of these events could not be correctly predicted and hence adequate precautions could not be taken to tackle it. This was because 'to the administrator as distinguished from the botanist the importance of this flowering of the bamboo lies in the fact that it is followed by a period of severe scarcity' which the former were called upon to deal with.[26]

After Hennikker's note, which was quite an exhaustive attempt to understand the phenomenon and bamboo flowering, the next report came from J.H. Lorrain, the missionary who had the experience of witnessing both the *mautam* and *thingtam* famine of 1910–11 and 1930–1. Lorrain had the advantage of closely working among the people rather than rely on official reports. Unlike Hennikker, Lorrain was convinced about the connection between bamboo fruits and fertility of the rats.[27] He was also sanguine about the correctness of the tribal prediction of the next bamboo flowering based on its periodicity. On the basis of this report he requested the government to take precaution for the next famine that was likely to occur in 1930–1. Both Hennikker and Lorraine agreed with the tribal prediction for the next famine and accordingly prepared a calendar of these events for the benefit of the officers as well as people.

For the colonial officials and scientists the flowering of bamboo was a mysterious event which was recorded again and again but no scientific explanation offered.[28] Among them were Baker,[29] Blanford,[30] Blatter,[31] Bradley,[32] Brandis[33] and Bourdillon[34] and in later times Father Santapu[35] and Chatterjee.[36] However none of them were aware of the phenomenon as it happened in the Mizo Hills. The people from the rest of India were also not allowed entry into these hill areas without official permission. The official explanation was that the tribals were wary of outsiders while the real reason was that the authorities wanted to keep the tribals far from the influence of nationalist movement. However the first synoptic survey explaining the bamboo flowering, which too did not take the Mizo Hills into cognizance, was available only in 1967. It was conducted and by Daniel Janzen.[37] Janzen regretted that the natural flowering of bamboo was being reduced into a unique event as natural bamboo forests were being denuded across the world.[38] During flowering individual aerial stems sometimes live for much less time than their species cycles and flower only at the end of the cycle when an inborn signal initiates the formation of inflorescences. Fruit development in a few species was also reported. The size and shape of bamboo fruits varied according to the species. There are 17 general and 22 species of bamboo which bear fruit.

Prediction of Famine

The colonial administrators were impressed by the veracity of the tribal prediction of famines. Aided by the Mizo elders, the colonial administrators draw up a calendar of famines which is given below. The year of actual occurrence as per the records, varied by one or two years only.

Nature of Famine	Year
Mautam	1860–1
Thingtam	1880–1
Mautam	1910–11
Thingtam	1930–1
Mautam	1960–1*
Thingtam	1977–8
Mautam	2007–8

Note: *The 1961 famine actually took place in 1957–9.

The Rat–Bamboo Connection

Already by 1925, the bamboos had started to flower, true to the prediction of an impending calamity in 1929, and signs of fear among the people became apparent. Soon there was a massive increase in the number of jungle rats.

There was no doubt that the famine was caused by rats invading the standing crops. What remained a mystery was the rapid multiplication of the rats after they had consumed the bamboo fruit containing seeds. Alexander Mackenzie, a colonial functionary, wrote in 1884, that 'the famine arose according to the concurrent testimony of all persons concerned, from the depredation of rats. In the previous season bamboos had seeded; the supply of food thus provided caused an immense increase in multiplication of rats'.[39] There was corroboration of this from missionaries working in the area, and like others they were also perplexed by the possible connections between the bamboo seed and the multiplication of rats.

The periodical flowering, seeding and dying down of certain species of bamboo all over these hills was followed last autumn by an enormous increase in the number of jungle rats . . . the connection between the flowering of bamboos and invasion of rats is a disputed point, but the theory which seems to be most satisfactory is that the bamboo fruits has the property of making the rats which eat it, extraordinarily prolific. Whatever may have been the cause directly, the bamboos has seeded and the rats begun to increase and swarm everywhere.[40]

In a letter to the administration, the Reverend Lorrain, one of the first Christian missionaries to the Mizo Hills, wrote,

It appears that the rats begin to get more than extraordinarily troublesome years before the simultaneous seeding of the *mau-thing* bamboo but as soon as the seeding was over they increased to such an extent that no human power could save the crops from their depredation.[41]

The official report on the Lushais prepared in 1888–9 reported that 'strange it may seem, the advent of rats appears to be a recognized calamity'.[42] In 1861–2 the *mautam* had affected Manipur also. Bamboo had begun to flower from 1857–8 itself and the population of the rats began to show a visible increase intimidating the tribals of the region. Colonel McCulloch who was a witness to the phenomena in Manipur in 1858, reported that the Kabui Nagas of Manipur on the south of the Cachar–Manipur Road were living in dread of a similar rat invasion due to bamboo flowering,

'Another calamity consists in the visits of the immense quantity of rats. These in their progress destroy everything before them. They nip down the standing corn, ascend the granaries, fill the houses and leave nothing behind them fit for human subsistence. Neither fire nor water stops the progress of the innumerable hosts.'[43] Although it was established that bamboo seeds had something to do with the increasing rat population, no one was sure of the explanation. Although the tribal were firm in their belief that it was the consumption of bamboo fruit that made the rats prolific, the administrators and missionaries had their own explanations.

The theory that gained credence was the tribal conviction. Nevertheless people, the administration and missionaries though were firm in their belief of the theory, they made no effort to establish its scientific basis.

THE TRAUMA

The third recorded famine during British rule was in 1912 and affected the region covering the Mizo Hills, Chittagong Hills and the Chin Hills areas under the jurisdiction of the Burmese government. This famine was of the *mautam* variety. An eyewitness account of the event goes as follows:

> We are in the midst of a famine up here. From Chittagong Hills tracts right away into Burma, the whole country has been overrun by a myriad of jungle rats, and nearly all the crops, which should have been reaped last November and December, were devoured. Since October, I have been travelling all over south Lushai Hills helping the government to relieve the distressed. Statistics have been collected showing the needs of every family as far as possible These have been met by having a supply of rice brought up to Demagiri. The people who are able bodied have to go to fetch it up, often many days journey across the mountains.
>
> Up to the present the wild yams and sago palms in the jungle have helped people to keep themselves alive, but once the yams have begun to sprout and will then be of no use for food until next cold season and the sago particularly grow only in certain localities. During the next few months we anticipate a time of great anxiety. If the people can only manage to tide over the rains, there is every hope that all will be well. The great fear is that

an epidemic may break out. We trust however that we may be spared such a calamity.

The rats having eaten up all the crops and every other thing available are now themselves dying out, apparently of starvation. There is therefore every reason to hope that there will not be enough left to devour the next harvest.[44]

Another missionary further reported the human tragedy as well. It was reported that the Mizos were traditional rat eaters. Captured rats would be used as food. But during the *mautam*, there was an abundant supply of rats.[45] Such an abundance perhaps diminished the utility. Moreover the dried rats would hardly make up for the loss of rice, which was the Mizo staple food. Some of the tribals who had rice left from the last harvest struggled to protect it from the invading rats. The unfortunate remainder, who constituted the majority, would search the forest for roots, jungle yams and other wild produce. Wild sago palm was collected from the forest, dried, pounded and then sifted. Its powder was then made into a kind of dumpling that was wrapped in a leaf and boiled. The resulting food was a very sticky insipid mass, full of gritty particles.[46] Others ate a kind of wild yam. It grew as a creeper, the upper part of which is inedible but lower down it is a long tuber, rich in starch, and resembling a potato in taste. The root was vertical and often very long, so to get the tuber, the tribals frequently had to dig to great depth in very hard soil. Tragic instances were reported about such searches. One heartbreaking scene recorded was that of a grown-up man sitting near one of the dug holes, large enough to admit a man, crying like a child because after toiling for hours for the root, he found his way blocked by a huge rock. At another site there was a widow with her baby on the back, working with all her feeble strength to extract the tuber. Often she would become so exhausted that she would lay down to rest, only to find insects crawling all over her and if she did not get out of the jungle before dusk, the wolves would devour her.[47]

On the other side of the frontier, the British government of Burma too had waged a great battle against the rodents and destroyed several thousands of them.[48]

Coping with Calamity: Defences against Rat-famine

Since bamboo flowering and the consequent rat-famine had been chasing the Mizos all along their history, they had developed their own mechanism to cope with it. Unfortunately, the Mizo tribes did not have a written language to record their struggle. The neighbouring people who had a written language did not know much about the happenings in the deep jungles and high mountains where the Mizos lived.

As is evident, though the British were the first to record it, such famines were not new to the Mizos. It is a part of their life and history. In fact, Mizo chroniclers conjectured that the previous tribe who inhabited the present Mizo land had vacated the hills due to the severity of these recurrent famines, which they failed to cope with. According to their calculations the Mizo tribes migrated to the present habitat in about AD 1724 from the land between Run Lui and Tiau rivers in Upper Burma where they lived during the period AD 1540–1723.[49] Retrospective calculations by the tribal elders demonstrate that a major famine was experienced in the area in AD 1719, which was just before the Mizos reached the area. Since then they have experienced the hardship of such rat-famines in AD 1737 and 1767. The Mizo oral historical testimonies recorded that during the last of these eighteenth century famines the remaining of the former occupants of these hills evacuated leaving it totally to the incoming Mizos.

Production and Transmission of Calamity Knowledge

The Mizos, like many tribals, had had a turbulent past full of struggles. But famine related to bamboo flowering or *mautam* as they called it, remained supreme in their history and consciousness. Thus if an elder Mizo was asked to look back he would count time saying 'before or after that great famine'.[50] The impact of the famine on the demographic structure of the tribe was such that if a youth survived a *mautam*, the tribals would be sure that he would live long, to see another *lanongateu* or the seeding of another kind of bamboo called *rangia* which does not attract rats. But the tribals confirmed that there were very few who saw two *mautams*.[51]

One of the major tribal defence against this natural calamity was the dissemination of the history of *mautam* to the community as well as to the younger generation through oral traditions. We have already seen the knowledge of bamboo flowering than the Mizos gathered and transmitted through generations. This acted as a major data-base for countering future calamities.

The experiences of these famines taught the Mizos that the phenomenon of bamboo flowering was cyclical and its periodicity was about 30 and 50 years respectively for the two species of bamboo. It enabled them to predict the the next bamboo flowering, the difference between the two species of the bamboo flowering which they called *mau* and *thing* and analyse their causes and impacts. They could see that both varieties created an immense increase in jungle rats, which then resulted in food shortage. Although they did not have the science to exactly know what caused the increase in rat population they could correctly correlate the connection between bamboo seed and enhancement of the fertility rate amongst the rodents. Not just the rodents, they did not fail to notice that bamboo flowering was picnic time for a number of animals. A number of insects like the locusts and caterpillars of all variety increase enormously. But this increase is equally fatal to the agricultural vegetation of the tribals as they eat up the entire field. But birds, guinea fowl, chicken and pigs also love the fruit and increase phenomenally. This established the aphrodisiac and fertility enhancing quality of the seeds. But the tribe could not experiment it on their domestic animals and thus benefit from its free supply, as they did not have many domestic animals. A few number of pet chickens, *gayal*, and pigs that they had, would be finished in the beginning of the food shortage season through community feast and rituals to push back the famine involving animal sacrifice. However, the situation provided a grand opportunity for hunting, as the animals would roam open and freely during these times. Initially the tribe would take advantage of the situation by feasting on easy-hunts but gradually that would create further problems due to which they had to abandon the activity. The tribe hunts communally which means all the hunters who were warriors as well would be away in the jungle leaving the village vulnerable to invasion by the neighbouring enemy villages. It could be an open invitation to the enemy villages that were equally short for food to plunder and loot, or to the distant enemies, which were not affected by the bamboo flowering to invade and subjugate them. Hence frequent hunting would not be a very good proposition. Moreover, the meat of domestic animals was not enough to substitute their staple food grain. Gradually they would become averse to regular intake of meat. This entire information was gathered from the tribal elders by colonial functionaries and missionaries which were then systematized for the scientific community.

Method of Agriculture in Times of Crisis

Even though such famines struck once in about 30 or 50 years, its impact on the Mizos was profound. It devastated them completely. The Mizos were basically a nomadic people who kept on migrating from East Asia towards South-East Asia because of inter-tribal rivalry, search for new settlements, invasion by powerful tribes, the gradual decline in fertility of the soil due to *jhum*, absorption of the farmlands by jungles, water scarcity during dry season as well as other serious natural calamities. As nomadic people they hardly had any asset except for domesticated animals and slaves. Their mode of production was primitive which included slash and burn cultivation, hunting and food gathering, and raid and plunder of foothill villages. Although agriculture is the mainstay of the people it was in a rudimentary stage of development. Like rest of the people, of South-East Asia, rice was the staple food of the Mizos. But the technology of wet rice cultivation was unknown to them. Even the knowledge of terrace cultivation was absent. They produced dry rice in hill slopes by clearing the jungle. The varieties of rice produced were *magh*, *singtang* and *buman*. The latter was cultivated only occasionally.

The heavy rainfall that the region received helped the rice to grow. With this method however rice could not be produced sufficiently. In fact, rice was a luxury consumed occasionally. For most of the year it was maize that was the staple diet along with other vegetables like cucumber and yam. A contemporary witness reported,

They have a most primitive way also of cultivation . . . there is very little order to these cultivations. Indian corn, millet, cucumbers, melons, pumpkins and the calabash are all mixed up together and grow as they will but along with these spring up the weeds very speedily, far quicker indeed than the true grain, and the process of weeding their cultivations soon commences. . . . At the end of July the maize crop is ready to harvest and it is to this crop that the Lakhers [a sub-tribe of the Mizo] look for help to enable them to get through the year without starvation. Their land being not very fertile, they are unable to grow sufficient rice to last them the whole of the twelve months. Towards the end of October the first fruits of the rice harvest is picked and during November and beginning of December as a rule the whole of rice crop is harvested while for several months past they have been eating the fruit of the cucumber, the pumpkin and various other vegetables, including the cultivated yam and the '*Bia*'—a species of arum lily bulb which practically takes the place of the potato in the country and is excellent eating although if too much of it is indulged in one is liable to suffer from violent pains in the abdomen.[52]

People often saved rice from a earlier year because *mautam* was expected. Some families had a crop of millet (*kaka*) as it could be preserved for a long time.[53] Tapioca and Colacasia were also cultivated as extra crops through *jhum* as rats did not eat them. The most effective method to combat the famine was to cultivate an extra batch of grain in quick succession to the normal crop whenever there were indications of an imminent famine and prepared an extra cautious storage for it so that when the rodents devoured the normal harvest in the famine year, people could fall back upon the other crop.[54] But the normal harvest of the Mizos was hardly adequate due to its primitive nature. Such poor farming technology produced barely enough for the full year for a family. 'The method of agriculture is wasteful and extravagant . . . the

Lushai cuts down perhaps fives acres of tree or bamboo jungle and when this has become thoroughly dry in the months of March and April, he burns the wreckage to fertilize the land. Rice is sown in May, broadcast or dibbled on the burned hillside, preferably during soft falling rain, and after two or three weeding during the monsoon, the crop is reaped in December after running the constant risks of damage by drought, insects, wind, storm, excessive and uncontrollable weeds.'[55] If there are plain fields available, the Mizos would try cultivating it for wet rice where water is retained throughout the growth of paddy. This was done only in western part of the Mizo Hills where the inhabitants were aware of the technology. This was due to the idea that rats usually did not venture into the water to eat paddy. But all these measures were of little help against the attack of millions of rodents. In fact, even in normal times the harvest of the tribals were not free from the invasion of wild pigs, rats, bird, deer and insects against which they had very little defence.[56] Moreover most of the families were often required to render compulsory labour services to the farmlands of the chiefs leaving them with no time to work in their own lands. Hence the question of storage for crisis did not even arise. It was only the chiefs who had stored grains for calamities mainly because they had the largest share of land and besides their slaves the community members also rendered labour services in a priority basis for the cultivation of their farmlands. This is the reason a number of distressed families would offer themselves as *inpuisung* (people who due to poverty, food shortage, sickness or distress surrender as lifelong slaves against food and shelter) to chiefs. Such an economy is not inherently equipped to withstand any calamity creating food shortage. Hence even a mildest crisis could cripple the economy leaving them with no other option but to *pem*—the Mizo word for migration—to a food sufficient region.

Destroy the Cause

One sure way of guarding against the famine was to destroy the cause—the rats. During the famine of 1930–1 a massive number of rats were killed by the villagers according to administrative reports. Preceding the famine, by 1925, the bamboos had already started to flower, true to the prediction of an impending calamity in 1929, and signs of fear among the people became apparent. Soon there was a massive increase in the number of jungle rats. The people had already begun to destroy them. In December 1924, 45,000 to 50,000 rats were killed in Aizawl sub-division alone. This was the story of all the famines that had been recorded so far. Taking cue from the people, the colonial administration subsequently had officially launched a campaign for mass destruction of rats from 1925 onwards. During the rat-famine of 1911–12 which affected the Burmese side of the Mizo population also, the Government of Burma organized a great battle against the rodents and destroyed scores of thousand of them.[57] In this war against the rats the people, the administrative machinery and the church combined to fight the menace.

There were several species of rats in the Mizo country. Among them were the jungle rats and the house rats. The difference between the two was that a house rat was brown all over while the jungle rat had a snow-white belly. Among the jungle rats there was the black rat that could kill large fowls in a very short space of time and carry away domestic chicken in great numbers. The second species was the bamboo rats, which lived entirely on the roots and was a much larger animal, generally the size of a young rabbit. With its long incisor teeth it could cut through the toughest of bamboo roots. These rats were coloured but on rare occasions white.[58] Jungle rats were eaten while house rats were not. There was no way house rats could be prevented but for the jungle rats a trap was invented. A long basket of bamboo latticework was made, just sufficiently wide enough in diameter to admit the animal passing into it. At one end it was closed and the other end open. The open end was placed into the burrow hole of the animal and supported by sticks of wood underneath and securely fastened. The rats or even porcupines on coming to the exit of its hole would force its way through, into the basket finding no other outlet. Once inside the basket it was impossible for the animal to turn back. The latticework was plaited in such a way that on its attempt to turn round, it would contract and hold the body of the animal more firmly making it impossible for the animal to move until the hunter reached the spot.[59] Such traps would be made in hundreds during a rat plague and hundreds of them would be killed.

During the plague of rats one straightforward way of controlling their population was to eat them. Initially they would be killed and dried to preserve the flesh for future use but the enormous supply would create an aversion towards it. It was reported that the Mizos were traditional rat eaters. Trapped rats would be dried over the fire and then eaten as meat. But during the *mautam*, there was an abundant supply of rats. They lost interest in rat-meat and longed for their staple food-rice.[60] On the initiative of the administration, the tribals set and reset rat traps in their fields. Individual farmers could trap as many as 500 rats in a single night.[61]

As the agricultural consumables became rare, the community would fall back on their forest resources. The forest generally is bountiful but after the bamboo fruits had been exhausted the rats would devour all the fruits and underground roots of other plants. The plantain vegetation was widespread and an item that was spared by the rats. But what was spared by the marauding rats would be finished by the monkeys as their other food item were exhausted by the rodents. The tribal had not yet developed banana plantation. Only item that was left untouched was the wild sago palm and a root called wild yam. The first survive because the rats find them unpalatable and the other lie too deep below the earth. The forest products were enough to see them through a minor crisis like the *thingtam* famine. But in a devastating *mautam* famine the Mizos found themselves helpless. The marginal tribals would be starving from the very beginning of the famine. Such families would be dependent on the community for food and were among the first to surrender to the chief as *inpuisung*. These families would also migrate to the plains and beg even before the community had decided in favour of it.

There was also a particular kind of earth that resembled pipe clay in consistency but was slightly harder and of a greyish nature, which the Lakher tribal often ate during scarcity and pregnancy in women.[62] They claimed that it had the power of sustaining men for 36 hours or more when food was not procurable or subjected to great privation.

As far as the more practical part of the anti-famine preparation was concerned, the tribe made huge baskets with covers to store the foodgrains to protect them from the invading rat population. They also begin to cut the bamboos to construct new houses as after the flowering the plants would die and there would be a scarcity of bamboo stick for such construction.

Transform into a Community of Merchants

The Mizo were not traditional traders. They would often go to the foothill markets to exchange their forest products for scarce goods from plains. They included salt, iron, wooden implements and other such items. But during the crisis of rat-famine the entire village would turn into a community of traders. They would procure several forest products in much more proportion than normal times and descend to the foothill plains and exchange these goods for rice and such staple food items. Such period generally saw increasing plains–hills interaction. This was also a period of festivity among the plains people as they could easily sell of their surplus agricultural products and get forest products and even cash money from the tribals. The Mizo Hills saw two consequtive famines, in 1861 and 1881 during which there was record trade and mercantile activity between the hills and the plains. During these years there was enormous amount of rubber tapped from the rubber trees growing wild in the Mizo Hills which were then brought down to the foothill markets and sold. By this time the bordering plains were the colonized areas of the British. With the British entered a number of Marwari traders and shopkeepers who were linked to the Calcutta market. The Marwari traders purchased this bounty of rubber cheap and supplied them to Calcutta. This growing demand of rubber by the Marwari merchants enthused the Mizos so much that they procured huge amount of this forest product and brought them down to the plains. The easy cash enabled them to purchase their food requirements. In fact, the British records lamented that due to the over-tapping of rubber during the 1861–81 famines the entire rubber forest in Mizo Hills died prematurely and when the British actually entered the hills as the sovereign of the area in 1898 there was no trace of any rubber forests in the Mizo Hills.

Moving to the Land of the Plenty

The foothill plains where the Mizos migrated were generally the area where they practised their raids, kidnapping and head-hunting. But the tribals were sure there would not be any retributive campaign on them taking advantage of their helplessness. This was because the over the years the hills–plains interaction had developed into an institution based on mutual interdependence. The tribal did raid the plains occasionally and carry off people alive as well as heads of the dead, which created a fear psychosis among the people of that area. Yet these marauders were no strangers to the people of the foothills, as during the rest of the year the tribal would come down as friendly hillmen who would bring the forest products like cotton, rubber, herbs, ivory, honey and such products and exchange them for products like iron tools, salt, clothes, rice and other related consumables. Generally they would descend from the hills on the market day through the gateway called *dawr* or *duar*. In the Cachar and Sylhet districts of Assam and Bengal respectively, still a number of foothill areas, which are known by names, which has a *duar* prefixed or suffixed to it. As such market days, which were weekly or bi-weekly events used to be a festive occasion for both the hillmen as well as plainsmen. There was a history of mutual interdependence that developed into an institution. One reason why the plainsmen were not vindictive to the Mizos was that they were really on the periphery of the state boundary—either Bengal or Cachar. These states were not very concerned about the plight of the people on the frontier and most of the times it ignored the atrocities committed by the Mizo head-hunters on the plainsmen. As result these marginal subjects were at the mercy of the tribal plunderers without any military aid from their respective states and were left to defend themselves. Hence they could not afford to take advantage of the helpless situation of the tribals and risk deterioration of the relationship. The tribals would go back to the mountains, anyway, when the crisis blew over. The plainsmen would buy the products of the Mizos brought down from the hills, provide them food for work and even supply food items in charity. This enabled the Mizos to cope with major food crisis resulting from famine. However such migration en masse was a last resort when every other option had failed.

It also became obvious to the Mizos during *mautam* that despite the body of knowledge gathered round the phenomenon, there is precious little the Mizos could do about it. They were totally powerless against this natural calamity. In fact, repeated famines depleted their population considerably despite the continuous migration of their people from their place of previous settlement. Starvation death during famines until the next harvest was considerable. Due to the nature of tribal society, they neither deserted their community nor did any nuclear families migrate to other places. The Mizos therefore temporarily migrated to the plains of Cachar and Sylhet during the famines and lived on the charity of the plainsmen. They returned back to the hills only when the severity of the famine had subsided. In this process some smaller sub-tribes permanently settled in the foothills near the plains and never went back to the high hills. The Mizos settled near the Hailakandi district of Assam thus trace their migration to this area.[63] In fact, recent scholars detected a distinct migratory trend among the Mizos towards the plains of neighbouring Assam and Bengal, which was put to a halt by the advent of the British in the region who made conscious efforts to stop the migratory movement of Mizos and to confine them within the hills.[64] The regional historians conjecture that if such efforts were not made the major portion of the Mizos would have come down to settle in the plains of Cachar and Sylhet like the Kacha Nagas, Hmars and Brus and other such tribes. During such settlement some Mizo tribes even picked up the technology of settled rice cultivation in the foothills through terrace cultivation. Despite their history of head-hunting, there is no record of hostility between the tribes and plainsmen during the temporary migration to the plains due to famine hardships.

Rise of Communitarian Philosophy

The famine destroyed the community too. The Mizos were a communal tribe. All the property was owned communally and all the works related to farming were also done communally. The village comprised the community. The community ethos was the in-built defence mechanism against all struggles. 'The kind of life they traditionally lived had given them self-reliance. They had to fend for themselves, for their own village and nobody owed them a living. They fostered basic skills and simple virtues. Honesty, courage, self-discipline, mutual help, a readiness to organize and be organized were highly appreciated.'[65] In fact, the Mizo concept of *Tlawmngaihna* (roughly rendered as service to the community) is the essence of their communal life.

> A system of community obligation existed under the term *Tlawmngaihna,* implying public service. Crops of the sick would be tended by the strong, the chief's lands would be weeded as a mark of support, help would be given to rebuild houses accidentally burned down, warriors would volunteer when asked for, hunters would strive to be energetic in the chase and in general the good citizen was he who was foremost in meeting calls that were really necessary for the good of the whole village.[66]

A Mizo elder defined the philosophy of *Tlawmngaihna* as:

> *Tlawmngaihna* is the Mizo code of moral and good form. One cannot for example be regarded as *Tlawmngai* unless one is courteous, considerate, helpful, unselfish, courageous, industrious and ready to help others even at considerable inconvenience to himself. A *Tlawmngai* man or woman will always try to ensure that he or she does not stand in need of help form others and will try to surpass others in doing his or her ordinary daily tasks efficiently. We thus see that *Tlawmngaihna* embraces various types of activities and manifests itself in various forms, which can be summed up as 'Group over self' wherein self-sacrifice for the needs of others is the spontaneous outcome. A man who practises the precepts of *Tlawnmgaihna* is highly respected.[67]

But a calamity like famine crushed this humanitarian spirit by affecting every single person of the community. Generally in a crisis few are affected while the rest of the community shields them from it. But food shortages incapacitated individuals from serving others thereby destroying the spirit as well as the physical existence of the community unit by unit.

With the death of the community its political will too ceases. The polity of the Mizos was centred on the community and its chieftainship. With the disintegration of the community the polity begin to fall apart. The famine acted as a great leveller by reducing both the chief and the commoners to the same state of destitution. A chief who failed to provide succour to his people was no longer recognized as a chief. Once the chief was stripped off its position, his chieftainship ceased to work. At the most he could act as a leader in one of the tribes' great migration to the bounty-food area. But that was an extreme situation. Until then, the in-built coping mechanism of the tribe worked to combat a famine.

Exorcizing the Spirit

Though the Mizos had developed enough knowledge about the calamity, they were hopelessly unequipped to counter such natural disaster. The only thing they could do was escape to safer places to avoid the hardship. Their knowledge was empirical and not scientific. Hence they developed hardly any technology to escape the turmoil. Since it was severe in its impact and was natural in character they considered it a curse of the gods above. Hence they took to religion to evade the punishment. There were particular rites to ward off the curse.[68] Every year in the month of *Chhippa* (corresponding to June) they performed a ceremony called *Chakalai*, to drive out the evil spirit that caused the famine. The chief himself fixed the day of the ceremony. At noon on the fixed day the village crier would send the message that *Chakalai* would be performed that night. When night fell each householder threw out all the half-burnt firebrands from his house, shouting *Chaka si la, cha phao si la, hia kha thlong la, thla-tla tlongla*, which meant 'go away famine to Haka or Thlatla'. On this night the women would not weave or cook. At dawn rice was cooked with very little water and every one ate as much rice as he could and the whole day was declared *aoh* (spend fasting) for the entire village. This ritual was performed to drive away the famine from the household. The head of the household had previously borrowed grains and other products from his neighbours owing to some unfortunate circumstances that might have arisen in regard to his farming and according to Lakher Mizo custom his borrowed grains must be paid back in double the quantity of the actual borrowed amount. After the ceremony had been gone through, however a concession would be made. The borrower now had to refund only the same quantity of grain and other items, which he actually borrowed instead of double the amount. A man however cannot perform this ceremony simply to escape his liabilities of paying double 'the amount he has borrowed, but there were certain very rigid condition' laid down between themselves and only under very special circumstances and by the consent of these concerned and with their chief as this ceremony of the *Chakalai* could be performed.[69]

Practically all division of the Lushai–Kuki family believed in a spirit called *pathian* who was supposed to be the creator of everything and was a beneficent being but had however little concern with men.[70] Far more important to the average man were the numerous *huai* or demons who inhabited every stream, mountain and forest and to whom every illness and misfortune was attributed. The village *puithiam* (sorcerer) was supposed to know what demon was causing what trouble and what kind of ritual and sacrifice appease him. The Lushais (Mizo) entire life was spent in propitiating these spirits. It was believed that one such spirit caused the famine. The Mizos were not really nature worshippers; they did not worship sun or moon or any of the forces of nature. They appeased spirits or *huais* who were uniformly bad as they only bring calamity and suffering to men. During the epidemics that follow the famines, the Mizos signal the evacuation of the village as bad spirits had possessed the village. The sick were abandoned and people scatter, some families taking up their abode in the *jhum* huts, others building huts in the jungles. The neighbouring villages close their gates to all immigrants from the infected villages. To terrify the *huai*, who was supposed to be responsible for the epidemic, a gateway was built across the road leading to the stricken villages, on the sides of which rude figures of armed men made of straw with wooden spears and clappers were placed. A dog was sacrified and the skull was hung on the gateway. The more Hinduized Mizo sub-tribes in Tripura and Hailakandi district of Assam like the Riangs perform an elaborate Hindu ritual in front of a constructed idol of famine deity. During the famines the other Mizo tribes ceaselessly pray to *pathian* the saviour.

The famine also gave rise to myths. For example the explosion in the rodent population after eating the bamboo flower was explained by the villagers by saying that during *mautam* even vegetables like brinjals and insects like caterpillars turn into huge rats.[71] It was also believed that if bamboo fruits were fed to cows their milk production increased. Even cats and other domestic animals grew huge in size and reproduced more litter then they normally did.[72] It was also reported that rats grew as big as piglets during *mautam* and they were born of mother earth rather than rat-mothers.[73]

The bamboo flowering and famine were such integral part of the Mizo life and history that alongside myths, interesting legends were formed. They not only reflected the inextricability of famines in Mizo culture but also the community ethos and values. One such story was of the two orphan brothers Thanghou and Liandou. During one such famine the two brothers moved from place to place in search of food. Hunger and fatigue had taken its toll on them. On the brink of death and desperate for food, the two brothers at last reached a place which had not been affected by scarcity. They saw with bewilderment that a community of farmers had just completed their harvest and were carrying them home. With a lot of hope they ran towards them but on reaching found that the community had left with their harvest. They knew even after carrying all the harvest home some amount of grains would be left behind in the field. But alas! It was not their lucky day. All they found was a single grain of millet. A single grain of millet and two hungry stomachs! The brothers were desperate for food and each wanted to eat that single grain. But brotherly love triumphed over starvation and they decided to share that single millet and save their lives for another day. The story demonstrated that spirit of sharing which was so dominant in the communal life of the Mizos.

The other story was of a little girl brought up by her stepmother. During a famine the entire family of the girl, that included her father, stepmother, stepbrothers and stepsister moved from place to place in search of food. When other consumables were exhausted they entered the jungle and dug deep into the earth for *hakai* (a root resembling *yam*). But a day came when even the jungle *yam* had finished. All the family was left with just one last dinner after which the entire family would have to die of starvation. While the mother was cooking the last supper the little girl was getting desperate for food. She kept pestering her mother to hurry up and serve her the food. Enraged by the impatience of the girl, the stepmother hit the hand of the girl with the serving spoon severing her fingers from the hand. Hurt and tearful, the starving girl immediately turned into a bird and flew away. Since then she would fly from village to village and say *Ka hut, Ka hut* (Cuckoo, Cuckoo), which in Mizo meant, my hand, my hand. This was the story of how a Cuckoo bird was born which is generally seen in spring time—the period when bamboo flowers and famines stalked the hills of Mizoram.[74]

NOTES

1. The details of this attack and the British response are available in Suhas Chatterjee, *Mizoram Under British Rule*, Delhi: Mittal, 1985.
2. Lieutenant Governor's report quoted in C.E. Buckland, *Bengal Under Lieutenant Governors*, vol. 1, London, p. 180; cited in J.B. Bhattacharjee, *Cachar Under Brtish Rule in North East India*, Delhi: Radiant, 1977, p. 113n.

3. Reginald A. Lorrain, *5 Years in Unknown Jungles for God and Empire*, Liverpool: Lakher Pioneer Mission, 1912; rpt. Gauhati: Spectrum, 1988, p. 165.
4. Zairema, *Gods Miracle in Mizoram*, Aizawl: Synod, 1978, p. 2.
5. A.G. McCall, *Lushai Chrysalis*, London, 1949; rpt. Kolkata: Firma KLM, 1977, p. 55.
6. Ibid.
7. Colonel E.B. Elly, *Military Report on the Chin-Lushai Country*, Simla, Government of India, 1893; rpt. Aizawl: Tribal Research Centre, Government of Mizoram, 1980, pp. 14–15.
8. *General Administration Report, Cachar District*, Shillong: Assam Secretariat Press, 1860–1.
9. 'Report of Colonel McCulloch', in H. Brown, *The Lushais 1888–9*, Shillong: Govt. of Assam, 1890; rpt. Aizawl: Govt. of Mizoram, 1978, p. 101.
10. F.C. Hennikker, *Maotam in Lusai: Notes Compiled in July 1912*, Maidstone, 1930, p. 6. Hennikker Papers, Box no. 10, Cambridge: Centre for South Asian Studies, University of Cambridge.
11. 'Report of Colonel McCulloch', in Brown, op. cit.
12. J.W. Bradley, 'Flowering of Kija Thuang Bamboo (*Bambusa Polymorpha*) in Irome Division, Burma', *Indian Forester*, 25, 1899, pp. 1–25.
13. C. Rokhuma, *What is Anti-Famine Organisation Doing*, Aizawl: Author, 1988, p. 97.
14. Ibid.
15. Hennikker, op. cit., p. 6.
16. D. Brandis, *Indian Trees*, p. 662, cited in Hennikker, op. cit., p. 1.
17. C.E. Low, *Gazeteer of Balaghat District* (*Central Provinces*), cited in Hennikker, op. cit.
18. Hennikker, op. cit.
19. *Foreign Political Proceedings* (*FPP*), Pol.A, August 1882, nos. 88–91.
20. Alexander Mackenzie, *History of the Relations of Government with Hill Tribes of North Eastern Frontier of Bengal*, Calcutta, 1884; rpt. as the *North East Frontier of India*, Delhi: Mittal, 1994, p. iii.
21. *General Administration Report, Cachar District*, Shillong: Govt. of Assam, 1890.
22. Rev. J.H. Lorrain, to the Superintendent of Lushai Hills, 27 January 1924, DC(A), No. 16/24–5 (1924–5), Mizoram State Archives Records.
23. Hennikker, op. cit.
24. Ibid., p. 3.
25. Ibid., p. 1.
26. Ibid.

27. J.H. Lorraine, 'Report on the 1911–12 Famine in the Lushai Hills', in *Annual Report of the Baptist Mission Society*, South Lushai Hills, Assam, 1912.
28. Ibid
29. E.C.S. Baker, 'The Game Birds of India, Burma and Ceylon', in *Journal of Bombay Natural History Society*, 24, 1916, pp. 201–3; cited in Vishwanathan, *The Carnival of Science*, Delhi: Oxford University Press, 1997, pp. 204–5.
30. H.R. Blanford, 'Note on Operations in Bamboo Flowered Areas in Kartha Division', *Indian Forester*, 44, 1918, pp. 550–60; cited in Vishwanathan, op. cit.
31. E. Blatter, 'The Flowering of the Bamboo' (Part I), *Journal of Bombay Natural History Society*, 33, 1929, pp. 899–992; cited in Vishwanathan, op. cit.
32. Bradley, op. cit.
33. Brandis, 'Biological Notes on Indian Bamboos', *Indian Forester*, 25, 1899, pp. 25–50; cited in Vishwanathan, op. cit.
34. T.F. Bourdillon, 'Seeding of the Bamboo Theory', *Indian Forester*, 21, 1895, pp. 228–9; cited in Vishwanathan, op. cit.
35. H. Santapu, 'The Flowering of Strobilantnes', *Journal of Bombay Natural History Society*, 44, 1944, pp. 605–6; cited in Vishwanathan, op. cit.
36. D. Chatterjee, 'Bamboo Fruits', *Journal of Bombay Natural History Society*, 57, 1960, pp. 451–3; cited in Vishwanathan, op. cit.
37. D. Janzen, 'Why do Bamboos Wait so Long to Flower?', *Annual Review of Ecology and Systematics*, 7, 1976, pp. 347–91; cited in Vishwanathan, op. cit.
38. Vishwanathan, op. cit., p. 206.
39. Mackenzie, op. cit., pp. 325–6.
40. McCall, op. cit.
41. Rev. J.H. Lorrain to the Superintendent of Lushai Hills, 27 January 1925, DC (A), No. 16/24–5, 1924–25, Mizoram State Archives Records.
42. Brown, op. cit.
43. 'Report of Colonel McCulloch', cited in Brown, op. cit.
44. J.H. Lorrain, *Revenue Proceedings*, no. 2071, 1913, dated 19 March 1912, Assam Secretariat Records.
45. *Annual Report of the Baptist Mission Society*, South Lushai Hills, Assam, 1912.
46. Ibid.
47. Ibid.
48. Ibid.

49. Rokhuma, *What is the Anti-Famine Campaign Organisation Doing?*, Aizawl: Author, 1988, p. 96.
50. Ibid.
51. Ibid.
52. Lorrain, op. cit., p. 116.
53. Interview of Upa Chalhnuna, *Mautam: The Experiences*, Aizawl: Govt. of Mizoram, 2006, p. 10.
54. Hennikker, *Maotam in Lusai*: *Notes Compiled in July 1912*, Maidstone, 1930, p. 6. Hennikker Papers, Box no. 10, Cambridge: Centre for South Asian Studies, University of Cambridge, *Note on Maotam*, Shillong, 1930, p. 4.
55. McCall, *Lushai Chrysalis,* London, 1949; rpt., Kolkata: Firma KLM, 1977, pp. 31–2.
56. Ibid.
57. *Annual Report of the Baptist Mission Society*, South Lushai Hills, Assam, 1912.
58. Ibid., p. 190.
59. Ibid.
60. *Annual Report of the Baptist Mission Society*, South Lushai Hills, Assam, 1912.
61. Ibid.
62. Lorrain, op. cit., p. 167.
63. Oral testimony collected from Katlichera, Assam, January 2002.
64. V. Ruata Rengsi, 'Pre-Colonial Technology of the Mizos', unpublished Ph.D. thesis, North-Eastern Hill University, Shillong, 1998.
65. J. Meirion Lloyd, *History of the Church in Mizoram (Harvest in the Hills)*, Aizawl: Synod, 1991, p. 4.
66. McCall, op. cit., pp. 97–8.
67. Sangliana, a Mizo social-anthropologist quoted in Rokhuma, op. cit.
68. N.E. Parry, *The Lakhers*, Calcutta: Macmillan, 1932; rpt. Kolkata: Firma KLM, 1988, pp. 272–3.
69. Lorrain, op. cit., pp. 119–20.
70. J. Shakespear, *The Lushai–Kuki Clans*, London: Macmillan, 1921; rpt. Kolkata: Firma KLM, 2002, pp. 62–79.
71. Rokhuma, op. cit., pp. 137–8.
72. Ibid.
73. Ibid.
74. Collected from oral sources.

CHAPTER 3

Famine as a Site for Politics of Paternalism

The British described the conquest of tribal north-east India—either Naga or Mizo—spanning over a period of about a century as 'one long, sickening story of open insults and defiance, bold outrages and cold-blooded murders on the one side and long suffering forbearance, forgiveness, concession and unlooked favour on the other'.[1] It is obvious that the former part of the description was ascribed to the tribal and the latter to the British. It is not surprising considering that the British had already blamed the tribals themselves for provoking the British to conquer them. 'It should be first promised that for the annexation of territory, the Naga [or the Mizo] themselves were responsible. . . . We only occupied the hills after a bitter experience extending over many years, which clearly showed that the annexation was the only way of preventing raids upon our villages. It was impossible for any *civilized* power to acquiesce in the perpetual harrying of its border folk.'[2] The contest between the 'civilized' and 'uncivilized' had begun.

It was an extension of the colonial theory that the Indian empire was built in 'a fit of absent mindedness'. Thus the 'civilized power' of the West, which was going through its most crucial phase of expansion in India during 1840–60, went on to say that 'it was not desired to extend our rule into the interior' as the hills were not 'a land of flowing milk and honey, no glittering outcrops to raise thought of mineral wealth, no telling indications of reservoirs of endless oil'.[3] The conquest became necessary in view of 'incursions' and to 'protect the lowland'. The resolve to take possession of the hills was to 'reclaim its inhabitants from savagery'.

and when 'a footing in hills had once been obtained further territorial expansion because almost inevitable'.[4] In fact, a decade later too, when the Mizos objected to pay revenue to the British, the British officer reminded them 'you forced us to occupy your hills; we had no wish to come up here but you would raid our villages so we had to come and now you have got to bear as much of the cost of occupation as possible'.[5]

As can be seen, the British had already adopted a self-defined, self-righteous position as far as their confrontations with the tribals were concerned. The tribals were viewed as 'uncivilized' and 'savage' and their raids on the plains were considered 'incursions' into the British territory. It logically followed that the British despatch punitive expeditions to bring the aggressors to book and halt the tribal raids. Thus, as per as the British were concerned, the consequent conquest of the hills was the just from a moral and legal point of view. It was also often emphasized that for the occupation of their hills, the tribals themselves were responsible—not the colonialists. But the story of the annexation of the Mizo Hills was not as simple as it was presented, as can be seen below.

The Confrontation

The Battle of Plassey (1757) had sealed the fate of Bengal. It had virtually handed over the province to the East India Company. In the year 1760, the *de jure* Nawab of Bengal ceded Chittagong, the south-eastern part of the province to the Company and trade was opened up with the adjacent and intensely forested hill tracts inhabited by tribes such as the Chakma, Magh, Tipperra, Mros, Khyeng, Kumi among others. It also brought them in contact with various Lushai tribes who lived in the bordering hills. This was the time the British were informed of the periodic head hunting and kidnapping raids perpetrated by these hills tribes on the plains people who were now British subjects. Since these were independent areas, the responsibility of controlling the raiders was given to a Manipur chieftain for the period 1834 to 1841. In 1842, a series of raids was unleashed on the inhabitants of the British

Indian districts of Aracan in the south and Sylhet in the north which prompted the British to take control of the situation in their own hands. This followed the protracted story of raids and counter-raids between the two adversaries.

The Kachubari raiders (1844) were punished with a strong hand.[6] The leader of the incursion Lalchukla—a Paite chief was chased, captured, tried and deported. But from 1849 onwards, the raids into the British foothills extending from Manipur in the east to Sylhet in the west intensified. There was a violent raid on Cachar in November 1849. The Government of Bengal despatched Frederick Lister with a force of Sylhet Light Infantry, which attacked the large village of Ngura and released four hundred captives. In his report, Lister advocated that to control the tribes a force of not less than 3,000 men were required. '500–1,000 of these would be required for keeping open the road from Cachar and protecting the various Depots . . . the remainder for carrying on operations in the country. A portion of this force ought to be European.'[7] Such extensive operations in difficult terrain with European troops at that stage was something which the Government of Bengal was not inclined. It therefore preferred to establish outposts in the frontier and raise small militia comprising martial tribes like the Kukis as a counter to the Mizos. The government also noticed that there were intense inter-tribal wars inside the hills, due to which the weaker tribes often offered themselves as subjects of the British against protection from their rivals as seen in the 1850s. Throughout the 1850s occasional raids into British territory continued with a severe one in 1862 in the villages of Sylhet frontier under the leadership of the powerful Lushai chief, Suakpuilala. The Bengal government wanted to send an expedition against the outrage but the Cachar authorities preferred negotiation as they had close contact with Suakpuilala. The negotiations resulted in an agreement of friendship between the two parties but the extension of tea gardens to the Lushai territory alarmed the tribals. They considered it a trespass into their territory and a prelude to the eventual conquest of the hills. Their chief Vonpilal, on behalf of others lodged an official objection to the British advance into tribal territory. Captain Steward, Superintendent of Cachar, tried to convince Vonpilal that the tea plantation would benefit the Lushais.[8] This failed to convince the Lushais. The younger chiefs construed the advance as an encroachment on their rights and united to offer determined resistance. A series of violent raids were committed in Cachar, Manipur and Tripura in November–December 1860. On 10 January 1869, raid under the leadership of Suakpuilala plundered the garden of Lowarbund and killed several labourers. On 14 January, another attack was launched on the Monierkhal Tea Estate and destroyed its building.

The situation in the foothills became alarming. The protracted atrocities, the loss of life and property of the British subjects as well as European planters compelled the Governor-General-in-Council to decide on punitive expeditions again. The local authorities suggested that

> raids should be met by condign punishment, in the shape of a military occupation of the raiders village during as long period as possible, the seizure of crops and stored grain and the forced submission of their chiefs; after that try the steady endeavour of the frontier officers to influence them and to promote trade and finally try a system of frontier posts combined with a line of road running north and south from Cachar frontier to that of Chittagong.[9]

Thereafter, then the above principles became the guideline for British intervention in the Mizo Hills. As can be seen later, famine neutralized the need to seize the grains of the tribal. In fact, the famine was appropriated to enter and control the tribe from within as per the above policy. A three-column expedition was sent which had to retreat due to incessant rain and trouble in the procurement of supply. The upset authorities declined to send another expedition in view of the failure of the preceding one. The Governor-General-in-Council instead decided on the appointment of an officer to effectively supervise the affairs of the tribe who would also regularly meet the chiefs and take engagements of good conduct. This officer would be placed between Cachar and Chittagong Hill Tracts. In addition the frontier outposts were to be strengthened and measures taken to prevent the supply of arms and ammunitions to the tribes from across the border.

Meanwhile a deputation of the tribal chiefs also met the Deputy Commissioner of Cachar, Edgar at Silchar to renew the amicable relationship. Edgar sought to avail this opportunity to place British relations with the Mizos on a more satisfactory footing. Accordingly Edgar took a tour of the Lushai Hills in December 1869, met Suakpuilala and other chiefs, granted them charters defining their duties and terms of relations with the government, fixed the boundary between the Cachar and Mizo territory and identified spots for outposts near the boundary.[10] The government too approved the charters granted to the chiefs specifying the contributions on which they would be left undisturbed in their territory, the levy of tolls by them on the people going up for trade with Mizos, the settlement of the villages along the frontier and the opening of two roads one from Manierkhal to Bawngkon in Mizo Hills and the other from Duarband to the Rengti Range.[11]

But the hopes of the authorities were shattered when extensive raids along the border of Cachar, Manipur and Tripura were committed again during December 1870. In Cachar, tea gardens were again the prime targets. In January 1871, they raided the gardens in the Ainarkhal in Hailakandi and killed 25 persons and took 37 as prisoners. Next day they plundered the Alexandrapur Tea Estate, killed its manager and kidnapped his six-year old daughter. On the same day they also attacked Katlicherra and killed five labourers. Three days later they attacked Dharnikhal and fought a pitched battle for two days with the armed police posted there. Next the plantations in Nagdigram were attacked and a number of sepoys killed. In February, the Jalnacherra Tea Estate was attacked and four persons killed. The matter reached such a pass that the government had to do something decisive. A military expedition was sent in two columns from Cachar and Chittagong respectively under General Browhow and General Brouchier. The expedition was truly massive in terms of its size, arms strength and resolve. The expedition achieved spectacular success as most of the chiefs were compelled to submit to British authority, release the captives and surrender their guns. But the local authorities felt that unless the success was followed by invidious policy, the effect would peter out. Edgar suggested that the success be followed by 'humanizing' efforts through extension of trade and barter and education.[12] T.H. Lewin, the Deputy Commisoner of Chittagong Hill District on the other hand advocated complete subjugation of the tribes.[13] But the government preferred non-interference in the internal affairs of tribals while keeping a strict vigil on their activities along the frontiers.[14] This policy led to the legislation of Bengal Eastern Frontier Regulation Act 1873, by which no British subjects were allowed to go inside tribal territory without permission. The idea was to stop the planters and plainsmen of whom the tribal were suspicious, to encroach the interiors thereby provoking a tribal raid on the frontier. One of the British officers posted in the area quite candidly admitted, 'We are quite powerless in preventing such raids and would be equally so if we had ten times our present force. In the kind of jungle which covers the hills a band of savages can always slip by unobserved and the effect of our police guards is almost entirely moral.'[15] He suggested that apart from exemplary punishment, efforts should be made 'to open up this unknown tract and to make its inhabitants feel that they were surrounded on all sides by a single government with a single aim and a single method of working'.[16]

Although from 1873, trade relations between the tribals and plainsmen began to grow and the size of foothill markets at Sonai, Tipaimukh and Changsil expanded, the British began to interfere in the inter-tribal feuds and succession rivalry. In October 1874, the Sailo chief, Lalngura sought British help against his enemy Benkuiya, the Howlong chief. Again, despite the opposition from the Lieutenant-Governor, the local administration interfered in the hostility between the eastern Mizo chiefs and Laljeika and Lalngura in September 1881. The same period witnessed violent battles, arising out of *jhum* lands, between the western and eastern chiefs. The anarchy was compounded by the death of the powerful western chief, Suakpuilala on whom the British depended heavily, and whose demise also brought a war of succession among his descendents. The growing hostility and violent feuds both among the tribes and within a tribe made the

British apprehend trouble as all these were sure to have repercussions on the frontier and raids, murders and kidnapping were anticipated. As the British authorities were preparing to counter the imminent turmoil, they confronted a new situation. Instead of raiders and murderers they found a host of destitute tribal refugees who were arriving in great numbers into the British territory. Although, they were apprehensive it became apparent that there was some crisis inside tribal country.

Advent of the Famine

The earliest indication of distress in the interior was the immigration of some 80 families from the village of Khalkam led by the eldest son of Suakpuilala.[17] This was followed by waves of other migrants—first the eastern chiefs with their followers then the western chiefs. The largest contingent of refugees came down the valley of Dhaleshwari, past Jalnacherra. This arrival not only alarmed the administration, it created fear among the plantation labourers who took them as raiders. The managers of the tea gardens, which lay on the route of their migration, also became cautious. But, soon it was found that they were not only peacefully inclined but were also distressed. They did not intend to raid but were anxious to earn a living by selling bamboo and forest produce and even by labour or begging. It was reported that,

> The chiefs sold out their ivory jewellery and other valuables for the sake of food. They exchanged their guns and other arms for food. They lost their numbers due to plague, pestilence. Their jhooms were exhausted and even rubber, which offered ample means of subsistence, was failing. They had no means to purchase articles such as salt, tobacco, etc. In short they were reduced to a state of destitution.[18]

This was unthinkable. In an unexpected turn of events a formidable enemy was willing to make an abject capitulation. They were ready to exchange their only weapons for food. The British had once concluded that the only way to subdue the Mizos was to starve them to submission by destroying their harvest and blocking their supply.[19] In fact, they had actually experimented with this strategy without much success. The British were curious to know what had caused the new shortage that succeeded in doing what the colonial machinery could not.

The authorities on enquiry found out that the cause of their distress was, indeed shortage of food in the hills.[20] In fact, the first news regarding scarcity of food was received in October 1881. On 17 October, some 80 households of the Rangte clan, all subjects of Khalkam migrated to Cachar and settled near Dharmakhal Tea Estate. They stated that they had been impelled by want of food and other causes to leave their homes. The news was then speedily confirmed by the arrival of men from villages of Khalkam, Poiboi, Lalhai and Chunglena who had travelled together. Shortly afterwards other groups of refugees arrived from Senvung village and the *Punjis* (Ramlet) of Ratanpur and Lalkunga.

About 400 subjects of Khalkam migrants settled in a Kuki village near Dharmakhal Tea Estate. By November, the tribal began to pour in through Tipaimukh and Jalnacherra. In the first week, 102 arrived from Servung and were allowed to reside in the Akai Kuki Punji in the Barak reserve. At the same time a few Mizos appeared in the Hailakandi valley. A few days later another numerous body of Mizos rafted down the Dhaleshwari River. On 10 November, information was received that a very large party of 2000–3000, including women were nearing the Jalnacherra outpost.

The cause of this large-scale migration was reported to be 'destruction of food crops by rats'. The Mizos explained that after certain intervals of years, the bamboo plants flower and a sort of seed is formed in them. The existence of this rich supply of bamboo fruits attracts rats in swarms and the rats procreate and grow in number. The rats then spread and consume everything edible, sparing neither paddy crops, *kachu* (*yam*) nor even cotton seeds. This resulted in the devastation of entire good stores and an acute shortage of foodstuff which began to cause starvation deaths amongst the tribals. So grim was the crisis that the Mizos were compelled to seek help from their arch enemy—the white British. In other words, it was a famine caused by the depredation of rats.[21]

Although the Mizos were victims of a natural calamity, their arrival in such great numbers in the plains caused apprehension. The plainsmen viewed them as fierce and savage raiders, who kill without reason, hunt heads for rites and kidnap men, women and children without compunction. Hence, 'it was a most difficult matter to persuade the garden coolies that these men were not bent on raiding but had come down merely to buy rice. Everything was done to calm the general apprehension and at one time it was almost thought necessary to increase the force at the outposts'.[22]

Not just the labourers, even the European planters were fearful and hence against allowing the tribes to pour into British territory. The raid of the Alexandrapur Tea Estate and the murder and kidnap of the Winchesters in 1871 were still fresh in memory. The European manager of Katlicherra Tea Estate immediately wrote to the Deputy Commissioner about the danger posed by the advent of these tribals into the plains. The Deputy Commissioner reported that there was no reason to apprehend any danger as the Mizos were not likely to do any harm and that they were there only in search of food. To reassure the manager, it added that an order has been issued to dispossess all those passing through Jalnacherra of their arms. But the manager wrote:

> I am in receipt of your letter of the 9th instant and am not at all surprised at your urging yourself with the belief that there is no danger to be apprehended from the Lushais in continuation of that policy which was carried out during Naga disturbance and ended in Baladhan tragedy, which was a disgrace to civil administration of this district. We will pass over the danger that is likely to happen to any Lushais at the hand of garden coolies; but about the Lushais not doing anything to give any offence is a matter in which I differ with you completely. The very presence of these starving vagrants is an offence and is a breach of section 518 of Indian Penal Code also when they wander about your bungalow and factory at sundown, where there is valuable property, or go into coolies houses, beg rice and demand food and I think it is time the government moved.
>
> You say owing to failure of crops they are in danger of famine, but is there any reason why the Hailakandi valley should be caused such uneasiness and by the sale of forest produce which they bring down, is the

famine to be averted because if so their want must be few indeed. Why the journey down must take some seven days and quite as many to go back again and if you think the sale of about 500 bamboos will keep a Lushai family until the next years crops' are ripe you must be credulous indeed.

Again you say you are issuing strict orders to constable at Jalnacherra that no Lushai will be allowed to pass the outpost armed but valuable and energetic a servant as the head constable may be I do not see how he is to prevent other Lushais coming down through the jungle with arms for their countrymen who come by water.

I demand in the interest of my employees that you as head of the district shall do something to allay the fear and inconvenience that the presence of these wild people cause to the natives in these parts and that they shall not be allowed to take possessions of our bazaars and buildings or whatever suits them to the inclusion of their owners.[23]

Although the district administration was trying to allay the fears of the plantation managers, they themselves were not very sure.[24] It was apparent that the tribals were distressed and starving but that there was a famine in the hills could not be ascertained. The administration was not sure about the reason of distress: was the migration of the hill folk due to famine or inter-clan violence or for that matter, a ploy to commit more depradations. Even if it was a scarcity, the danger was more as certain clans were likely to raid the plains and tea estate for food items. It was also feared that marauding raider might follow the distressed migrants. To confirm the situation the district administration decided to send Rai Bahadur Hari Charan Sharma, the Special Extra Assistant Commissioner, who was thoroughly acquainted with the Mizo tribals, their customs, language and who was personally known to the chiefs with whom he had many interviews along with Mr. Place, the Sub-Divisional Officer of Hailakandi to Tipaimukh and Guturmukh to make an on-the-spot enquiry into the situation.

The Rai Bahadur left on his tour of enquiry among the Western chiefs, on 11 November 1881 and confirmed that a rat-famine resulting from bamboo flowering had actually hit the hill areas and the hillmen were in acute food scarcity. In the middle of November, *mantris* (ministers of chiefs or elderly men) arrived from the villages of Lalsavenga, Sailengpui, Banaitang and

Lengpunga with requests for food. They reported that people were subsisting chiefly on roots. That rats had not only destroyed the standing crops but devoured the grain in the granaries. The officer had interviews with the hillmen who were living in temporary huts above Jhalnacherra. He found that they belonged to the Pibuk Rani's (chieftainess) *punji* (village). After the death of their Rani some had taken refuge with Baniatang, others with Ratanpui and such chiefs of other villages. In consequence of the scarcity, they had taken up their abode below Damcherra with an adopted son of Ratanpui, said to be the natural son of late Suakpuilala by a Kuki woman. These groups expressed a desire to *jhum* on government land but they were directed to abstain from doing so and to remain south from Bhairabcherra which was the boundary between British territory and the Mizo (Lushai) country. At Pakuacherra, there was another encampment of 100 households who were anxiously waiting for the arrival of relief from the plains. They were delighted to see the Rai Bahadur visit them. According to them they had been without food for two days. Proceeding further up the river, he met about 100 tribals coming down on 15 bamboo rafts. They were hungry and the children among them were the worst sufferers. These people were on their way to join their commanders at Pakhuamukh. Their immediate wants were food. In Khutbul-Chouramukh at Ratanpui's *punji* there were about 600 Mizos who were quite destitute, mostly surviving on roots, leaves and fruits. The chiefs declared that the scarcity was not confined to their village but that the *punjis* of all Lushai chieftains were similarly placed. The Howlong and Pois too had been affected. The Rai Bahadur reported that there was a general and widespread calamity and unless assistance were supplied there would be large scale mortality from starvation. The chiefs earnestly requested the administration to direct the traders to make advances of grain to the affected, repayable after harvesting of the next *jhum* crops. The chief wanted certain trade marts to be opened inside the hills and relief in terms of loans of grains be provided by the government. The chiefs agreed that they themselves would be responsible for the repayment of loans.

The Making of the Famine Policy

The authorities in Silchar, headquarters of Cachar district confirmed that the subsistence economy of the Mizos had been crushed by the famine. As food stocks depleted, starvation deaths followed. Warriors turned into beggars and many exchanged their guns and arms for food. Some deserted or died while others begun to migrate to the hitherto enemy territory. There was serious apprehension that such a state of affairs could give rise to renewed raids and plunder on British territory to obtain food. The new Deputy Commissioner of Cachar J. Knox Wright received intelligence messages that the tribals were preparing for desperate raids on the Sylhet, Cachar and Tripura borders.[25] It was also reported that the older generation was against such misadventure in times of such crisis.[26] They were ready to come to the British and seek assistance in mitigating the distress. In fact, the impact of the famine was so great that rival clans gave up their age-old hostility and made temporary truce. Poiboi, the eastern Lushai chief, Kalkhom the western chief and Lalhai, son of Vonpilal met in a conference and decided to make strong endeavour to secure food from the British authorities in Cachar. Accordingly, Lushai elders (*upa*) met Knox Wright, informed him of the distress and sought help to tide over the trying time.

Relief in the form of food items were being already distributed to the refugees. It had to be now decided whether relief should be provided or taking advantage of the crisis the Mizo head-hunters should be subdued once for all. There were many considerations:

1. Should the tribal, who were hostile to the British, and as such an enemy should be shown any mercy during their time of distress.
2. If relief were to be provided, the tribals would soon tide over the famine and go back to their old ways of raid and plunder into the British territory.
3. If they were not provided relief they would continue to migrate to the plains causing uneasiness in the tea gardens. Both the tea labourers and the estate management were against the

permanent presence of such a large number of 'savage' tribesmen around them.

4. Subjugation of the tribals taking advantage of their helplessness required a larger policy decision regarding whether the British Indian territory was to be expanded into the hills or not.
5. Even if the decision veered towards conquering newer territories, past experience showed that conquering the tribes through war was out of question, as the hills were difficult terrain for modern warfare. It also had a large number of tribes, all of whom had to be conquered together if they were to be subjugated at all and this was not possible from a military point of view.

The local administration was convinced that conquest of all the tribes at one time was not possible. In fact, the British would rather have left the tribals alone, as long as they did not disturb British territory through their raids and aggression. But that too was much of an expectation. The only way was to earn their goodwill, generate a friendly relationship and provide them with necessary supplies so that they did not raid the plains. The overall issue at that point of time was to gradually make them dependent on the British, teach them ways of civilized behaviour and slowly get a foothold into the heart of the hills from where they could be easily monitored and controlled. Moreover, once an entry into the hills had been obtained, the British could get a fair idea of the geography and terrain of the hills and most important of all, try to open roads throughout the Lushai Hills so that tribal movements could be monitored. The trusted lieutenant of the local administration, Rai Bahadur Hari Charan Sharma, on whose views the local authorities had immense faith, also emphasized on demonstrating the benevolence of the British to the 'ignorant and savage Lushais'.[27] He felt that the lives of these primitive tribals could only be saved if the government consented to act with munificence. He was certain that such acts would compel the hostile tribals to see the friendly face of the British Raj and cease hostility.

J. Wright Knox, the Deputy Commissioner of Cachar was himself of the same view. He was located in the hotbed of the events from which he could actually see the fierce opposition of the Mizo and was convinced of the difficulty in completely subjugating the head-hunters. Moreover he was presiding over an area which was growing prosperous by the day through the flourishing tea industry. He knew that unless these head-hunting tribals were appeased, the tea gardens at the foothills of the tribal habitat were not secure. Although a section of the British officials and the tea planters themselves were against any move to provide succour to the tribals, Knox Wright was sanguine that a policy of conciliation was bound to bear fruit. He understood that whatever might be the strength of the British it was not easy to win a war against an enemy who refused to fight an outright confrontational war. The tribes would strike at their convenience and run away to the deep interiors and a counterattack would be difficult. In such a situation only a moral conquest was the option. Moreover, this was in pursuance of the British policy towards the tribes adopted in the 1870s. Knox Wright's ideas had a context. One of the earliest British officers to deal with the Lushai was Captain Lister who had unequivocally stated that the 'Lushais were a virile race, possessed a clear-cut culture pattern and capable of giving endless trouble at any time anywhere along the British Indian settlements bordering Lushai, unless subjugated once and for all.'[28] But years of punitive expedition and its repeated failures in completely subduing the tribes made the Indian government rethink the strategy. There was also a financial consideration. Each of the expeditions cost the army a huge sum without any tangible results. There was acute criticism from the home government as well as the English media. The Government of India therefore adopted a 'general policy of appeasement which really implied a determination to resist the financial costs of occupying hills. Directions were issued for operations but purely retaliatory measures were to be avoided as well as needless harm to innocent people. It was the government's wish to seek friendship, not conquest.'[29] Knox Wright saw that the crisis had provided the British providential opportunity to earn the gratitude of the hill people. What several years of war and genocide failed to earn could be achieved in weeks by coming to the aid of the tribals. Knox Wright decided to convince his superior of the judiciousness of the move. He wrote to the Chief Commissioner of Assam, emphasizing the crucial role 'famine mitigation' could play in normalizing the relationship with the tribals. He echoed the same hope in writing to his superior, 'It would be quite best a policy, apart from the call upon our humanity to secure the goodwill of these chiefs by affording help in their distress.'[30]

The Chief Commissioner based in Shillong contacted the Lieutenant-Governor of Bengal apprising him of the situation on the Indo–Lushai border and the report containing the idea of Knox Wright on the way of dealing with the situation. The famine had not spared the Mizos living in the southern hills (Chittagong Hill Tracts) which were under the direct jurisdiction of the Bengal government. It was also monitoring the developments and dealing with the problem of famine-affected Mizo refugees. The distressed Mizos of the Bengal frontier had also appealed to the authorities at Chittagong for help. The Lieutenant-Governor of Bengal Sir Bailey, agreed with the perception of the Assam administration that assistance to the Mizos in times of their distress was an opportunity to develop friendly relationship for future benefits.[31] He accordingly authorized the local officers to arrange for the storage of grains at Demagiri and other convenient stations. He however instructed the Deputy Commissioner of Chittagong Hill Tracts to keep direct touch with the Lushai chiefs and not allow intermediaries in his dealings to ensure visibility of the administration as the benefactor. The Deputy Commissioner requested the chiefs to allow stationing of a *karbari* (trader) inside the hills. The *karbari* was provided with a subsistence allowance and quarters to live. His job was to look after the distribution of the relief material and also report on the changing situation. The Chief Commissioner at Shillong, as well as the authorities in Calcutta, agreed with Knox Wright and gave the local administration at Cachar and Chittagong the go ahead with relief works. The Chief Commissioner of Assam, however, had a word of caution,

All relief should be provided but nothing should be done to pauperize the refugees or create the impression that the able bodied among them could get food without working for it. Nor should the means be taken to relieve them as such as will tend to saddle us with the care of them for a period longer than necessary. It is not desirable either, that they should settle down in Cachar district and it is hoped that when the pinch is over, they will return to place from where they came.[32]

It was decided that relief would be provided to the applicant chiefs on condition that the relief be treated as loans and repaid within a year. It would ensure that the chief remain committed to the British even after the famine was over and did not create any trouble. On 1 November 1882 the local authorities submitted an application to the government for sanctioning initial expenditure of Rs. 2,000 only. It was calculated that the population of the villages where scarcity was the greatest amount to 9,000 and the total cost of relief would be Rs. 3,000 It was realized that it was not possible to recover the entire amount advanced to the chiefs, but it was hoped that the chiefs would repay at least half of it. But the local administration was not much disturbed by the economics of famine-relief. They looked at the potential results of philanthropy. In the words of Knox Wright, 'It was considered *politics*, however to assist them even with the certainty of loss. It was felt to be much to be deprecated that the power of the Hillman with whom we happened at that time to be on excellent terms should be broken and it was thought that no pain should be spared to prevent the desertion of villages.'[33]

The Relief Operation

On the basis of the report of the Rai Bahadur, the Government of India made an immediate sanction of Rs. 50,000 for relief work.[34] An estimate of the requirement of rice was calculated: 'The Northern Lushais number more or less 30,000 souls and if two children be counted as one adult man the number would be 20,000 so that, if half a seer of rice is to be given per head daily, 20,500 maunds are needed for the three months.'[35] The actual estimate, which also had a rough estimate of the population before the first census in the Lushai Hills, is as follows:

Zones	No. of Persons
North Lushai Hills	30,000
South Lushai Hills	15,000
Total Population of Lushai Hills	45,000
Rice Consuming Population in North Lushai Hills (children excluded)	30,000 10,000 20,000
Rice Consuming Population in South Lushai Hills (children excluded)	15,000 5,000 10,000
Daily Intake of rice in North Lushai Hills per day (.5 seer per head) 20,000 × .5 seers	10,000 seers
Daily Intake of rice in South Lushai Hills per day (.5 per head) 10,000 × .5 seers	5,000 seers
Total consumption of rice per head per day (excluding rice beer)	15,000 seers 375 maunds (approximately 350 quintals)
Total consumption of rice per head per day (excluding rice beer) 350 × 365 quintals	1,277,50 quintals

At his instance only the authorities in Shillong issued orders to the *mirasdars* (*mirasdars* were revenue officials of the pre-colonial Cachar state who were retained by the colonial administration in the same capacity with the same designation) and other well-to-do cultivators of Sylhet and Cachar districts to collect rice from their tenants for distribution to the famine affected people.[36] The proposals were submitted to erect two government store-houses, one at Tipaimukh on the bank of river Barak, and the other at Guturmukh on the bank of river Barak, the two major ports and traditional trade centres through which the Lushai Hills could be penetrated. The other such traditional foothill markets were Sonai Bazar on the bank of Sonai River and Changsil Bazaar on the bank of Dhaleswari. These were strategic points for the British as well as the Lushais as they were the entry and exit points to and from the hills. The suffix *mukh* meant opening. The importance of these marts were stated by the first Superintendent of Lushai Hills himself,

> The history of our dealing with the Lushai chiefs is to a great extent interwoven with that of the markets which our traders keep up in Lushai land and it will perhaps be best to begin the subject of our relationship with the Lushais by a brief narrative of these events connected with each of these bazaars. These are the Tipaimukh Bazaar on the Barak, where the Tipai falls into it, the Sonai Bazar on the Sonai River and the Changsil Bazaar on the Dhaleshwari River.[37]

The British administration in Cachar always wanted to take advantage of these traditional centres of exchange between the hills and plains and develop them further. The famine of 1881 provided the British with the opportunity to develop and turn these foothill marts into centres of tribal dependence on the plains. After constructing store houses at these strategic points, the government stocked them with 1,000 maunds of paddy and 1,000 maunds of rice. The suggestion that a subscription in rice or money might be raised in this district did not meet with approval. It now became necessary to formulate a policy, and accordingly certain rules of guidance were laid down. The principles, which guided these rules, were threefold:

1. That, the loss of the rice crop in a hill district, can never imply the same loss of the means of subsistence which it implies in the plains: this doctrine, the Famine Commission laid down when they asserted that hill districts enjoy a practical immunity from famine in its severest forms.
2. That if hillmen were in want of food they should, where practicable, give labour in return for it.
3. That, if possible, government stores of grain should not be opened, but traders be encouraged and assisted to open the stores.[38]

The object of government was to provide all needful relief: at the same time, care was taken to avoid all steps that might tend to pauperize the refugees, or to create the impression that the able bodied among them would get food without working for it. It was particularly necessary to prevent any impression gaining ground that the administration was ready to keep them one moment longer than was actually necessary. The government made it clear that it was not at all desirable that any large number of the migrants should settle down in the district, and so make way for the northwards advance of the Pois and other southern clans. This advance was going on steadily. The *punjis* of Suakpuilala Khalkam, Ratanpuia, Minthang, and others were now considerably to the north of the positions which they occupied a few years ago. In fact, the pre-colonial movement of the nomadic tribes of the Mizo towards the Cachar plains had stopped with the advent of the British and the fierce encounter between the two. As a consequence of the famine, the movement began again. A large section of different tribes migrated to the plains and were indicating that they were about to permanently settle there. The colonial administration was aware of this and did not want it to happen. The authorities were convinced that the presence of head-hunting tribals around tea plantation areas would pose a serious threat. Hence it desired that the Mizos should leave the plains as quickly as possible and wherever possible their movement should be stopped and their distress should be tackled inside the hills. The British also prepared a list of existing villages in the Lushai Hills along with the number of households and people it contained as can be seen in Table 3.1.

Table 3.1 also mentioned the number of men and women each village had but it was in manuscript form and has become too illegible to read except for the total of 28,800. It also had an interesting remark under the 'remark' column. It stated, 'the first eight villages may be assumed to belong the eastern confederacy. Lalhai and Darkang will remain neutral, if possible, if any quarrel between the two principal sections of the Lushais (broke out).'[39]

It must not be inferred that all the measures for relief were adopted at the very beginning. They were introduced gradually,

TABLE 3.1: VILLAGE-WISE HOUSE/POPULATION DATA

Name of Chiefs	No. of Houses	No. of men and women	Remarks
Chunglenba	100	illegible*	
Poiboi	400	"	
Lalruma	200	"	
Lalcheri	100	"	
Lalhenga	100	"	
Bangtewa	250	"	
Lengkam	450	"	
Lalbura	400	"	
Lalhai	400	"	
Darkay	100	"	
Khalkom	500	"	
Rowapunji	50	"	
Thangula	150	"	
Sukpilal	300	"	
Thangbang	200	"	
Lengbang	300	"	
Darmangpui	200	"	
Banaitangi	300	"	
Lalkhunga	200	"	
Mentang	100	"	
Ratanpui	250	"	
Dausuma/Son Lotang	100	"	
Lalvunga/lunglena	500	"	
Lalkhuma, Sanglura	200	"	
Sailengpui	1,050	"	
Jarak	300	"	
Total	7,200	28,800	

Note: *Could not be deciphered from the record.

as occasion required. The authorities did not restrict themselves to the care of the immigrants only; it was necessary to do something for those who did not migrate to the plains.

The two government depots, as mentioned earlier, were intended to store paddy, and not rice. It was still found necessary to keep a stock of rice as well, not only in case of accidents, but also for the requirements of the more distant villages to which the transport of paddy would be somewhat difficult and inconvenient. It was not meant to sell the paddy. Those who had purchasing power were adviced to buy paddy from the traders; for others, the government paddy was stored to give out as a loan, on the security of the chiefs and headmen, to those who brought a certificate from their chiefs that they could not purchase. It was not expected, of course, that written certificates would be furnished, as the art of writing was wholly unknown. To meet the difficulty, seals bearing the names of the chiefs were prepared and given to them with directions to stamp their names on small bits of paper, which they were to give to anyone who was wholly unable to purchase for himself. These slips would be kept as vouchers by the depot-keeper. Each slip was to indicate one mound of paddy; but this system was subsequently found inconvenient, and at the request of those concerned each slip was declared to a rupee worth. The indebtedness in rupees could thus be easily ascertained by the chiefs, although little prospect was seen of recovery by the chiefs from their people of advances made to them. Although little hope was entertained of the government ever getting back what it advanced, still it was considered advisable to keep up the system of guarantee, because direct relief without any guarantee whatever, was likely to incur the danger of having the whole population of the hills dependent on the administration and the Lushais who actually possessed property would vie with the poorest in the community in applying for gratuitous relief. If, on the other hand, the authorities insisted on a guarantee from the chiefs and the headmen, they would probably exert their influence as far as possible to keep down demands to a reasonable low limit. The question of the total remission of the debt was reserved for the subsequent consideration of government.[41]

In the event of the traders' shops not being sufficient, instructions were issued that the government depots should be opened for purchases as well as advances. The price of the paddy advanced to the hill men was fixed at cost price in Hailakandi, the government bearing the whole cost of carriage and establishment. Moreover, a limit was placed on the profits of traders; they were prohibited from charging more than 4 *annas* per maund in excess of cost price and carriage. To carry out these proposals, a further grant of Rs. 3,000 (in addition to the Rs. 2,000 first sanctioned) was placed at the disposal of the local authorities.

In December, the number of immigrant Lushais increased, and among those who visited the Deputy Commissioner was the chief Dansama, son of Darmangpui. This was the first occasion on which he had ever visited Cachar.[42] In February, 500 maunds of government paddy reached the Guturmukh storehouse and another 500 was on its way.

As the famine intensified, the administration sent its emissaries to visit the interiors of the hills to report on the conditions. The tour dairy of the officials show the visits made to the Lushai border and country in connection with famine:[43]

11th to 26th November, Rai Bahadur visited Lushai country.

23rd November Deputy Commissioner visited Damehara Camp.

14th and 15th November, Mr. Place a government official, went to Jhalnacherra and visited a camp of Lushais at Damcherra.

1st December, he visited Jhalnacherra.

26th and 27th December, he visited Jhalnacherra.

6th to 29th January, Rai Bahadur visited Changsil, Guturmukh, &c.

19th to 23th January, Mr. Place visited Guturmukh.

1st February, Chief Commissioner, with Personal Assistant and Mr. Place, visited Jhalnacherra.

15th February to 10th March, Rai Bahadur visited Tipaimukh.

26th and 27th February, Mr. Place visited Jhalnacherra and Lushai Camp.

7th to 12th April, Mr. Place visited Guturmukh with Rai Bahadur, who stayed some time longer.

29th and 30th April, Mr. Place visited Jhalnacherra.

8th and 9th May, Mr. Place visited Jhalnacherra.

The Cachar administration also maintained meticulous accounts of the expenses it incurred in famine relief. The following tables showing the total expenses up to 8 June incurred by government on account of measures taken for the relief of famine.[44]

	Expenditure		
	Rs.	*Annas*	*Paisa*
Purchase of paddy	1,035	13	3
Purchase of rice	46	9	0
Purchase of stationary	4	14	9
Boat hire	1,036	8	9
Miscellaneous	117	5	9
Total	2,241	3	6

Received from the Silchar *mahajans* on account of government stores advanced to them:

	Received		
	Rs.	*Annas*	*Paisa*
Paddy 1,000 maunds	1,016	8	0
Rice 16 maunds 8 seers	–	–	–
received from Rai Bahadur	25	2	0
Total	1,041	10	0

The balance of cost to government was Rs. 1,199-9 *anna*-6 *paisa*. Of this, a portion of which was received on account of advances to the chiefs amounting to 696 *kathi* maund and 8 *kathis*.

The paddy account up to 19 May was as follows:

Total amount purchased at	*Kathi* maunds	*Kathi*
Monierkhal	2,501	9
Mainadhar	2,324	10
Jhalnacherra	191	8
Total	5,016	27

The difference of Rs. 14-9 *annas* is owing to an excess quantity being found at Guturmukh. The weighing was wrong.

The whole quantity of 2,324 maunds 10 *kathi* which arrived at Guturmukh was consumed as follows:

	(Rs.)	*Annas*
Wages of Lushais for erecting godowns	40	5
For landing and storing	51	3
Total	91	8
Advances to shopkeepers	1,536	14
Advances to Lushais on security of chiefs	696	8
Total	2,324	10

In addition to this, article sent to Tipaimukh was only 50 maunds of paddy, at a cost of Rs. 30. The traders were found quite able to supply full demands. Of the 50 maunds were expended for the labour of constructing the godown, 20 maunds were sold to a shopkeeper for Rs. 21 (not yet paid).[45]

Famine in Bengal Frontier: Chittagong

The 1881 famine affected the southern Lushai and other tribes of the Bengal frontier as well. The affected tribes were the Howlong, Sailos, Thanglowas and Reang Tipperas. In other words unlike other famines, this time the entire population of the Lushai Hills, both north and south were affected by famine. Like the administration of Assam, the Government of Bengal also came forward in assisting the famine-stricken people. In pursuance of the general British policy, all the officers of the Government of Bengal were unanimous in rendering the help to the Lushais. The Lieutenant-Governor of Bengal, Sir Stuart Bailey, informed the Government of India that he had authorized the local officers to arrange for the storage of grains at Demagiri and other convenient stations. The administration of Bengal readily agreed to assist the Lushai chiefs since this could have the way for a friendly relationship in future. The Deputy Commissioner of Chittagong Hill Tracts was instructed to stay in direct touch with the Lushai chiefs and not through any intermediary chiefs. The Deputy Commissioner requested the Howlong chiefs to keep a *karbari* at his headquarters. The *karbari* was provided with a subsistence allowance and a quarter. The Deputy Commissioner made a spot survey of the famine areas and submitted a detailed report as to the needs of supplies. Gratuitous relief was sanctioned and arrangements were made for transportation of rice to the sale depots. In his report of 5 March 1882, E. Lowis, the Commissioner, Chittagong, submitted to the Government of Bengal that about 3,000 houses were to be fed and allowing five persons for each house, he calculated that another 14,000 maunds of rice in addition to 5,000 maunds already purchased would be required for the coming months till the time of the next harvest. Like the authorities of Cachar, the Chittagong authorities also established rice godowns for relief operations in the trade marts at Kasalong, Demagiri and Fiskisiva.

Food for Work

As per government policy, the local administration withdrew all duties at the forest toll stations on timber and other forest products to facilitate exchange. The administration instructed the local authorities to offer employment. Following it, the forest officials employed the willing Mizo refugees for cleaning of forest boundaries. The tea garden managers also used the available extra labour force for cleaning the jungles, cutting down the woods and extending boundaries of their estates. The Mizos, though not accustomed to hoeing or road making were however skilful in jungle clearing and accepted work readily when offered on high wages.

The Mizo labour was also used for the construction of roads. It is important to remember that the British had not yet conquered Mizo territory. Yet the famine was used by the British to construct roads within the Mizo territory in clear contravention of Mizo sovereignty. The objective was that after the famine these roads could be used to reach the Mizo interiors. The hungry immigrants did the job ungrudgingly. It was by using this labour force during the famine conditions of 1881–2 that the British completed the Silchar–Changsil and Demagiri–Kabalong Roads. To keep the supplies ready for relief, some more trade marts were also established along the roads, easily reachable from Cachar and Chittagong. In fact, Rai Bahadur Hari Charan Sharma, the colonial functionary and in-charge of relief operations in the hills covered the entire habitat of the Mizo. He travelled the hills and recorded their distance from each other and supplied the description of the less known paths, which were used by the Mizos to ambush British troops. These reports helped the British during the subsequent expeditions and eventual conquest of the hills.

Policing the Interior

In the 1870s, the British Government was facing a troublesome period *vis-à-vis* the incursions of the Lushais, in both the Cachar and Chittagong frontier. Frustrated at the failures of the colonial military in stopping the head hunting raids and incursions into British territory, it was trying different policies and methods to negotiate with the tribals: first a forward policy of subjugating the tribes and then a policy of non-interference. During the period of non-interference,

> The Government of India however would not hear of an expedition. It was averse on principle to move bodies of troops or police to effect reprisals for outrage or chastise offenders by following them into their hills. But it was willing to try the plan of direct management by a selected officer. Frontier posts were to be erected and frontier villages armed for defence. The Lushais in short were to be managed by love, while they had not yet learnt the respect and fear which when followed by forbearance, alone lead such savages to love.[46]

As a loyal officer, Major Edgar had experimented with the policy. The English media had of course found this 'mode of treating a savage and hostile people was a 'policy without a backbone—a limb and nerveless phantom not to be leant upon at all'.[47] But it could not be denied that a beginning was made in negotiating with the tribe. The media also recommended new policies in context of the changing circumstances:

> First we must open the country as far as possible by roads, not necessarily *macadamised turnpikes*, but broad, serviceable paths along which a body of troops or police can march with elephants. One such path driven through from Cachar to Chittagong would do more to civilize the Lushais than any other schemes suggested. The experience of all hill tribes teaches us this. . . . Roads are essential both to conciliation and to repression. Some officers are, we know opposed to them, on the ground that they would open out the tea gardens to the Lushais as well as the Lushai village to our police. But surely this is a very short-sighted line to take. With our roads we must have frontier posts; we must have armed parcels. If a raiding party did slip past, they ought always to be intercepted on their return. The dread of this would any way tend to prevent raids; and in fact a system of posts and patrol paths has secured the north of the Chittagong Hills from inroads these ten years past. On the side of Cachar our policy whether conciliatory or not must rest on a basis of paths, posts and patrols. In the face of the strong (but weak) determination of the government to allow this anomalous Lushai land, a mere strip between two British districts to continue independent, we can do no more.[48]

Indeed, famine had provided an opportunity to the British to intrude into the deep hills, construct roads and paths and consolidate the marts where the tribals visited regularly into institutions of dependence. The administration did not also want to waste this opportunity to strengthen itself in the interior as a guard against the future hostility of the tribes. The government all along its charity was concerned that once the famine was over the tribe would go back to the hills and would resume its raid and head hunting activities on the British territory. Hence this was the right opportunity to construct outposts and man them by armed police. The pretext would be to guard the government storehouses. Thus a force of one head constable and twelve men of the Frontier Police was detailed to guard the depots at Tipaimukh and Guturmukh. In addition, a *muharrir* (indigenous accountant) was provided to keep the accounts at the former place and a sub-inspector in the latter.[49] Such stations were utilized by the British to mediate the internal rivalry between the tribes as these were a source of danger to British territory. For example, the opportunity of scarcity at Tipaimukh was taken advantage of to settle two disputes of long standing: formerly the chiefs would not permit any shops to be built on the west bank of the Tipai. The Lushais are unable to swim, and could not trust themselves to water. It was therefore a matter of extreme inconvenience and danger for the inhabitants on the west bank to get their supplies. After much trouble Poiboi and Lengkhan were induced to withdraw their objections, and five shops erected on the west bank. Government storehouses served the starving people while the policemen stationed there allowed the colonial authorities to arm itself inside tribal territory. The visible intention of the government disturbed the eastern clans who this time began to exhibit an uneasy feeling. As already remarked, a small force of constables had been sent to protect the depot at Tipaimukh. The chiefs expressed a wish they might be recalled. Careful enquiries were made, and when it was ascertained that there was no need for a government depot there, that 4,000 maunds of rice were in stock, that the traders were fully able to supply all wants, that the scarcity in this direction was not so severe as in the west, that all the sick and weak had already left and gone to reside within British or Manipur territory, that those who remained were able to take care of themselves and tide over the present distress, it was decided in March to dispose of the government stores and withdraw the guard, keeping only one *muharrir* or head constable to sell the paddy in stock, keep accounts, give out passes for bamboos, exercise general supervision over the traders, and submit to headquarters regular reports as to the course of events. This was accordingly implemented reluctantly by the administration.

Activating Commerce

In January and February, the trade between the hills and Cachar increased considerably. It proved to be quite a festive occasion. As the plainsmen and the tea-labourer realized that the tribals were distressed and peacefully inclined the fear disappeared and there was friendly exchange between the two. Markets grew up around the place where the refugees had settled. The tribals reciprocated with equal enthusiasm. The purchasing power of the Lushais had not yet been exhausted. Indeed, some of them still found means to purchase fancy articles, such as beads, combs, looking glasses, etc. The tribal would go back to the hills and bring back rubber, ivory, elephant hides and canes. During the famine year the amount of rubber exported from North Lushai Hills was 1,000 maunds and the size of bamboo export was 4,25,000 units. The total money value of the exported good was Rs. 61,800.[50] 'The Lushais were not acquainted with the money value of rubber, and they merely exchanged a certain quantity of rubber for a certain quantity of rice' reported the Deputy Commissioner. The result was expected. During the famine period the Lushais overtapped the rubber plants to such an extent that the entire rubber forest was lost. Major M.O. Boyd, the next Deputy Commissioner of Cachar, returning from his tour of the North Lushai Hills in March 1883 reported, 'between Tipaimukh and Changsil bazaar I did not see a single rubber tree, though I have been told that rubber trees still exist further to the south'.[51]

By this time, all arrangements had been completed and government stores of paddy had been collected ready for any emergency. The encouragement and facilities held out to traders had enabled them to supply the wants of the Lushais, who were able to make payments with money.[52] Tables 3.2–3.7 show the statistics of trade with the Lushai country during the famine months. It also shows the number of traders and Lushais who travelled from the hills and back for trade purposes[53] as can be seen in Table 3.2.

Table 3.2 showed that a general export of 18,063 maunds of rice and 2,062 maunds of paddy, exclusive of the government stores, were despatched into Lushai country. On the other hand, Table 3.3 shows the quantity of rubber, wax and bamboo brought down from Lushai country on the incentives provided by the Cachar administration, by the Lushais.

TABLE 3.2: RICE AND PADDY EXPORTED TO LUSHAI COUNTRY (IN MAUNDS—ABOUT 39 KG)

	Monierkhal		Mainadhar		Jhalnacherra	
	Rice	Paddy	Rice	Paddy	Rice	Paddy
November	80	–	548	–	557	–
December	165	–	796	–	483	2
January	319	–	1,979	–	1,181	40
February	–	–	2,293	–	2,335	689
March	626	35	899	33	3,644	889
April	–	–	281	1	1,877	373
Total	1,190	35	6,796	34	10,077	1,993

TABLE 3.3: VOLUME OF TRADE IN BORDER MARKETS

	Monierkhal				Mainadhar				Jhalnacherra			
	Rubber		Wax	Bamboo	Rubber		Wax	Bamboo	Rubber		Wax	Bamboo
	Mds	Seers			Mds	Seers			Mds	Seers		
November	–	–	–	–	–	–	–	–	48	30	19¼	29,360
December	0	30	–	–	3	35	–	–	4	22	–	7,130
January	1	6	–	–	32	20	–	–	1	20	10	3,900
February	1	26	–	–	14	0	2	–	5	0	–	3,990
March	0	12	–	–	0	8	–	–	3	23	–	15,650
April	–	–	–	–	–	–	–	–	3	0	–	15,415
Total	3	34	–	–	50	23	2	–	66	15	29¼	65,445

TABLE 3.4: THE QUANTITY BROUGHT DOWN BY NON-TRIBAL TRADERS

	Monierkhal				Mainadhar					Jhanacherra				
	Rubber		Wax	Bamboo	Rubber		Wax		Bamboo	Rubber		Wax		Bamboo
	Mds	Seers			Mds	Seers	Mds	Seers		Mds	Seers			
November	–	–	–	28,610	18	0	5	0	22,500	16	10	–	–	140
December	–	–	–	24,100	51	5	3	25	75,000	42	18	1	15	–
January	4	34	–	160	95	20	3	0	10,000	12	20	–	–	–
February	0	15	–	–	156	12	2	36½	43,500	21	0	–	–	100
March	–	–	–	–	133	32	0	10	68,300	61	17	–	–	14,300
April	19	0	1	1,125	84	0	0	8½	26,350	145	0	–	–	45,875
Total	24	9	1	53,995	468	29	15	0	2,45,650	298	25	1	15	60,415

THE GRAND TOTAL THEREFORE

	Rubber		Wax		Bamboos
	Mds	Seers	Mds	Seers	
Monierkhal	28	3	1	0	53,995
Mainadhar	589	12	17	0	2,45,650
Jhalnacherra	365	0	2	4½	1,25,810

THE NUMBER OF LUSHAIS AND TRADERS WHO CROSSED OVER INTO CACHAR

	Lushais			Traders		
	Monierkhal	Mainadhar	Jhalnacherra	Monierkhal	Mainadhar	Jhalnacherra
November	–	21	–	20	38	43
December	–	276	1,435	Not known	17	17
January	198	684	–	Not known	Not known	Not known
February	80	317	658	Not known	228	10
March	22	75	1,164	2	111	22
April	–	3	902	2	51	10
Total	300	1,376	4,159	24	445	102

TABLE 3.5: THE NUMBER OF LUSHAIS AND TRADERS WHO WENT TO THE LUSHAI COUNTRY FROM THE CACHAR PLAINS

	Lushais			Traders		
	Monier-khal	Maina-dhar	Jhalna-cherra	Monier-khal	Maina-dhar	Jhalna-cherra
November	–	10	–	33	50	58
December	47	89	1,340	2	92	66
January	198	341	–	22	137	59
February	9	30	376	21	112	85
March	10	51	754	54	18	127
April	–	4	975	74	25	66
Total	264	525	3,445	206	434	461

TABLE 3.6: THE MONEY VALUE OF IMPORTS AND EXPORTS

	Imports (Rs.)	Exports (Rs.)
Monierkhal	5,428	4,431
Mainadhar	34,050	18,802
Jhalnacherra	22,341	19,462
Total	61,819	42,695

TABLE 3.7: VOLUME OF TRADE

	Mds	Seers	Value (Rs.)
Via Mainadhar	589	12	23,402
Via Jhalnacherra	565	0	14,105
Difference	24	12	9,297

The above tables show that 5,835 Lushais crossed over into Cachar, whilst 4,234 returned, so that 1,601 must have remained. Some of these tribals had settled in the district, but many were likely to return later in the month of May and June as the administration was against the idea of Lushais settling in Cachar plains. The number of non-tribal traders who visited Lushai country was, 1,101; of these, 571 returned, leaving 530 behind who were to return subsequently.

The administration astonishingly reported that the famine had actually benefited British territory. In its Annual Report, the Deputy Commissioner stated [The Import-Export Return] 'Show(ed) a profit in favour of Cachar of Rs. 19,124. The famine has therefore benefited this district very considerably. It will be observed that the profit has been greatest in the trade via Mainadhar. The cause of this is the large quantity of rubber that came that way. The average price is Rs. 40 a maund.'

This item alone very nearly accounts for the greater profit derived from the trade through the two places.[54]

As the Lushais had already demonstrated a great earnestness in cutting and selling bamboos and other forest produce, tracts were assigned to them by the administration in which they were permitted to cut, and sell bamboos. Furthermore, facilities were afforded for the sale and disposal. It was found that the royalty charged on all bamboos and rafts at the forest toll-station acted as a great impediment to the sale of forest produce. So long as prices remained the same there was no reason for any increase in demand, and the traders, therefore, had no need for purchasing large quantities. A remission of royalty lowered prices immediately and the demand increased. Having so much less to pay, the traders could afford to give the Lushais a better price, while the increased demand caused by the fall in prices enabled the Lushais to dispose off their produce very readily. Care was taken to notify the public of the temporary remission of duty on bamboos and at the same time every precaution was taken to see that the Lushais profited by the remission of the royalty. To secure the Lushais from imposition of further royalty, they were all informed of the facts. As the remission was not intended to affect other woodcutters, orders were passed that all rafts owned by Lushais should pass free; but subsequently, on a representation made by them to the effect that they were unable to manage rafts, and that some of them had been drowned in trying to navigate them, permission was given to them to sell their forest produce at the bazaars, and the traders who then purchased from them got a pass signed by the officer-in-charge, testifying that the purchase of the things had been made from Lushais. The certificate, on production at the toll-station, enabled the forest produce to pass free. The forest officer was also directed to employ as many Lushais as possible in cutting pathways through the reserves, and generally they were to be employed wherever possible. Communications were held with the planters, and it was ascertained in which tea gardens the managers were ready to give employment to the refugees. In this way a large number of Lushais found occupation in tea gardens. The administration reported that, 'the Lushais were unused to the hoe, but they did excellent work in cutting down forested tracts. Their behaviour was exemplary, and no complaints had been received as to their conduct. It was true that they were unable to understand ownership of vast tracts of forest, and it was occasionally difficult to make them understand that they must not cut down bamboos or other forest produce in private lands.' The planters, however, understood their case, and did not press matters. The ignorance of the Lushais in the matter of money led to such imposition of additional taxes levied on them by the Manipuris. The hillmen would crowd to a bazaar, and pay just double what the rice was worth. The traders, finding it more profitable to deal with them than with the coolies, reserved their stocks, and in one or two instances the coolies found themselves unable to buy rice in bazaars, and they then told the refugees that they were being cheated and what they ought to pay.

The Lushais themselves had by then learnt the profitability of famine-time business. It was now that the Lushais began to turn to profitable account, the distress prevailing among the Sailo Kukis (called Howlongs) between Lushai Hills and Chittagong. These Sailos lived nearer Changsil than the bazaars in Chittagong. Formerly, there was no opposition offered to their attending the

Changsil Bazaar, but the Lushais now saw their way to doing a little profitable business. They forbade access to Changsil to the Howlongs, while they themselves purchased rice and sold it at enhanced rates to the latter tribe in their *punjis*. In addition to their desire for gain, the Lushais were also no doubt influenced by the fear that the Howlongs would buy up much of the supplies, and so reduce the quantity available for themselves.[56]

A very satisfied administration at the development of the markets commented in its *General Administration Report*, 'Three Bazars have been established in recent years in the Lushai country beyond our border which are supplied with goods by native traders from Cachar. They are increasing gradually in size and importance but their growth in somewhat checked by the exactions imposed on the traders by the chiefs.'[57]

Famine Afflictions and White Men's Image

Towards the end of February, cholera broke out at Tipaimukh, and this added to the distress caused by the famine. This could not be tackled, as the tribesmen would not allow the 'white man' to enter the cholera-affected village. This was because the tribals believed that wherever a 'white man' entered, it was followed by cholera. Unfortunately, instances of cholera increased among the eastern clans, and caused a general dispersion among hill men. Chief Chunglena died from the disease, and his people went to another *punji*. Eventually some of the chiefs relented to allow the Western medicine to be distributed by the administration which saved a number of lives and weakened the prevalent belief regarding the connection between 'white men' and the disease. It was a huge reward for the famine mitigation efforts of the Englishmen.

Another infliction, however, was reserved for the eastern Lushais. In addition to the famine and cholera, they were subjected to a raid from the Paite Kukis of Manipur. In May, Khatarkhoi Punji, the abode of Chunglena Rani, not far from the bazaar, was attacked, and 25 people killed, many were wounded and some were carried off. The whole *punji* was destroyed by fire. This was followed by wholesale desertion, said to be the result of the intrigues set on foot by the Manipur Senapati, who was in the country during February and March. This was the time when *jhums* were being cultivated and the crops were progressing. The rice shops that were opened during the famine were being gradually closed, as the necessity for them diminished. But the government vigilance was not relaxed until the *magh* and *batong* paddy had been reaped fearing a similar food shortage due to such inter-tribal hostility and the consequent migration of farmers.[58] The assumption of the role of protector and arbiter in tribal affairs without annexation of Mizo territory was another famine-time achievement of the British.

There was a second difficulty as well. The traders complained of excessive taxation by the chiefs. This had been removed through a discussion with the latter they were in a distressed position due to the famine and could not refuse the demands of their benefactor and an amicable arrangement was effected. Instead of Rs. 10 in cash, 5-tobacco-leaves, and a basket of salt well pressed down (nominally 25 seers), which were payable monthly on account of each shop, the chiefs had now consented to accept only Rs. 5, with the tobacco-leaves and a basket of salt not pressed down; the difference of the pressing said to be 15 seers. The leading chiefs made a third concession: they had agreed to allow Chunglena the duties payable on two shops out of the 32 which then existed. From being an outsider, the British had reached the position of dictating terms to the chiefs even though they had not yet been conquered.

The Aftermath

In April, when Place and the Rai Bahadur made a tour in the west Lushai country, matters remained much as they were. Rai Bahadur reported that there was no extra cause for anxiety from the tribals or famine and trade continued briskly between Cachar and the Lushai bazaars. The scarcity still continued, though, and would continue until June or July, until the reaping of *jhum* crops, *magh* and *batong* paddy, but there were no signs of its increasing in intensity. The indigent had migrated to Cachar or Manipur, the able bodied found full occupation, some on tea gardens, some on government works, while hundreds were provided with employment by the traders in cutting bamboos and other forest produce. The chiefs were at peace among themselves. Towards the end of March the immigrants began to show signs of returning to their country and of cultivating their *jhum,* and the numbers who attended the bazaar diminished. In May the state of affairs had improved. The worst of the famine seemed to be over. The recent rains had caused an increase in forest produce (such as roots, leaves, bamboo shoots, etc.), which the Lushai consumed. As the season advanced the famine began to cease. With the improvement in the situation, the Lushais who had entered Cachar began to return in order to prepare their lands for the next *jhum* cycle.

The policy of coming to the aid of the Mizos earned their goodwill. In fact, the administration did not discriminate between friendly and hostile villages in the distribution of relief. By July 1882 the entire group of migrants had returned to the hills from Cachar. The entire Mizo population was overwhelmed by the 'kindness' the *sirkar* had shown. There was no denial of the fact that without large-scale liberal relief the famine would have taken a much heavier toll on the population. The Mizos were used to the calamity but this was the first major famine after their large-scale migration to the present site. In fact, such relief was also unprecedented for the Mizos and that too from an enemy. A triumphant Deputy Commissioner, who felt vindicated, reported,

> A feeling of confidence and gratitude prevailed among the western Clans: the fear and distrust of Europeans diminished to such an extent that the Assistant Commissioner of Hailakandi was invited in some cases to visit the punjis. This is noteworthy, for the Lushais are very jealous in excluding Europeans, and are very superstitious in believing that cholera always follows in their wakes.[59]

The famine however also exposed their weaknesses. In face of a natural calamity the Mizo war-making capacity was useless. The Mizo economy was so subsistence based that they hardly had any surplus for the rainy day. Moreover, technologically they were so poor that they could not tackle small creatures like rats. It also betrayed that they were dependent on the plains for all time—whether on a regular day or at times of crisis.

That the Mizos were still suspicious of the English was clear. During the famine the British constructed roads, paths and by-paths, trading outposts at strategic centres, placed loyal plainsmen in those trade centres and even posted frontier police in few locations. This did not escape the eyes of the clever Mizos. The chiefs who received aid from the British kept quiet about it but the chiefs from the far-east who did not suffer much from the famine ravages, deserted these postings quickly and even expressed their fear that this might lead to the eventual annexation of their territory. In fact, the British had to finally withdraw them. The western chiefs suffered most and they were grateful to the English for the assistance during the calamity. They reciprocated accordingly. Place, the Assistant Deputy Commissioner of Cachar on his visit was permitted to enter the *punji.* Such an invitation was never given to any Englishmen before as the tribals believed cholera follows the entry of white man. The courage to go against their age-old superstition was a significant metamorphosis.

The famine took place in 1881–2 and British relief continued up to January 1883 and by 1898 the conquest of the Mizo Hills had been completed. This was not a simple juxtaposition of dates. The British annexation of the Mizo Hills was dictated by the British tea plantation interests in the Surma–Barak Valley and the final decision to conquer was prompted by the British success against the Burmese in the Third Anglo–Burma War (1885–6) which resulted in the annexation of upper Burma. The famine of 1881 played a very important role in debilitating the Mizos and the relief operation strengthened the British strategic strength inside the hills. The famine ravaged the Mizos economy. About 15,000 people perished due to starvation. Among them were the warriors and chiefs. The guns and ammunitions had been sold or exchanged for food, and could never be recovered. The famine was followed by pestilence, plague and other such epidemics. A good number of fighting men migrated permanently to Manipur, Jampui hills of Tripura and Chittagong Hill Tracts. Rubber trees died of over-tapping and no articles were left to barter for salt, tobacco, thread, among other things. In short they were in a state of destitution even before the famine was over. J. Knox Wright, the Deputy Commissioner at Cachar, who had considerable knowledge of the eastern frontier and who advocated assisting the famine distressed with relief wanted the Government of India to take advantage of the plight of the Mizos. He felt that the time and opportunity was right for the British to strike. The famine had devastated the Mizos but it had earned the British the good-will of the people. Therefore instead of indirect control—which was the policy, outright annexation and assumption of direct control, as was the case in the Naga Hills was advocated. He said as much in a communication to the Chief Commissioner of Assam on 28 June 1884.

> The time has come when we must form more intimate relations with the Lushais. We must be prepared shortly to *Naga hillise* them. We must assume domination over them of sort or other, we must open out their country, we must provide for thorough communication between this and the Chittagong Hill Tracts, we must be in a position to know what is going beyond our borders, we must throw forward our present stockades and not leave miles and miles of valuable reserve to become the hunting ground of this Lushais. With decent protections secured to the traders, a great impetus would be given to the timber trade. Valuable land well suited for tea is to be found in the south. The rubber trade may be revived and it will be found remunerative to have plantations as soil and climate are well adapted to the growth of the plant. Above all the security that will be offered to our borders will well repay any cost involved in the scheme I have here briefly sketched.[60]

Although the opinion of Knox Wright was highly valued, the Chief Commissioner turned down this particular proposal. He considered the display of force would be better used to prevent further raids. The Governor-General-in-Council too shared the view of the Chief Commissioner and offered to sanction a force along the frontier with the twin objectives of making a demonstration of strength and at the same time inspecting the demarcated boundary. The Viceroy in fact did not share Knox Wrights views on the causes of the tribal raids and rejected the idea of complete subjugation as the tribals had calmed down after the famine. But Knox Wright proved right and a series of raids and plunder followed in quick succession. First the Sonai Bazaar was looted. Then, on 16 January 1885, a police boat was attacked and British sepoys were carried off. The authorities realized the importance of Wright's view and were concerned with the deterioration of the situation in the hills. The authorities at Calcutta were too busy elsewhere to think of the Mizo Hills. But the annexation of Upper Burma (1885) rendered the utility of Mizo Hills as a buffer between Burma and British India unnecessary. Therefore the docks were clear for the final absorption of Mizo Hills. This time, the Knox Wright policy of winning the 'goodwill' of the Mizos worked as the British received assistance from the friendly chiefs in conquering the hostile ones.

The Recurrence

In tune with Mizo predictions, the next big famine (*mautam*) stalked the hills again in 1911 and again in 1929. The report of the Lushai Hills for the year 1909–10 stated that the harvest was exceptional due to 'well distributed rainfall'. The harvest was better in Aizawl than Lungleh. As a result 'food stocks on the whole were in excess of normal'. Yet there was shortage of foodstuff. The official report states 'The partial failure of crops in 1910–11 as an indirect result of the flowering of bamboos was followed by serious scarcity all over the district. The effect of this flowering was to cause a tremendous increase in the number of rats, who destroyed all the crops.'[61] The bamboo plants flowered in 1910 and there were indications of rats by the end of March 1910. It was very painful to the farmers to see their bumper harvest devoured by the rats. The damage by rats continued throughout the year and even 1911 witnessed no respite. In June 1912 the maize and millet was about to be ripen when the rats destroyed the crop.

The crop of rice which should have been reaped about November-December 1911 was almost entirely destroyed by rats with the result that by January 1912 the stocks of grain began to give out and from then onwards it was incumbent on government to afford relief.The Lushai Hills were very difficult of access from the outer world. No railway cab be utilized to import rice quickly to the place where it is needed; wheeled traffic is out the question on hill paths and river communication is difficult and only available at three widely separated bases. There are no markets and even if the Lushai has cash they could not buy food without making a long journey of perhaps ten days or more to the plains. In Lushai Hills the means adopted by government to save people from what must otherwise have meant starvation for thousands has been to grant advances of rice by means of an order on traders who brough the rice up by river Sairang, Dema-giri or Tipaimukh.[62]

But this time the Mizos did not have to migrate to British territory for relief. They had been turned into British subjects themselves and the British had prevented them migrating from hills to the plains. All the migratory movements were stopped by the colonial state as it felt migration was one of the causes of inter-tribal rivalry and raids into the plains. All the *duars* (foothills) were sealed and the tribals were confined within their territory. An endeavour was made to encourage them to settle in a place permanently by facilitating all the amenities within that area. Special care was taken to halt the tribal migration towards the plains. The foothill plains were very important to the British as these were tea-growing areas. Any disturbance in the plantations could incite the media in England to launch a vigorous attack on the administration. Even the non-tribal visits to the interiors were prevented by promulgating the Bengal Eastern Frontier Regulation Act 1873 by which no non-Mizo, either European or Indian plainsmen could enter the Lushai Hills without written permission from the administration. This was a serious blow to the tribal economy as the Lushai were absolutely dependent on these exchanges with the plains through the *duar* markets. The closure of these *duars* exchanges implied that all the provisions including scarce commodities were to be procured and supplied to the hills by the administration. The administration was willing to do it through its traders but not risk allowing the tribals do business with the plainsmen. The issue of famine relief this time therefore had to be tackled within the hill region. Similarly providing relief to the Mizos were no more a ploy to 'earn the goodwill of a hostile tribe' and part of 'good politics' but a responsibility of the colonial administration. But given the record of the British in tackling famine situation elsewhere in India, notably Bengal, it is interesting to note how they reacted to the famines in the Mizo Hills. In the rest of India a famine was followed by the constituion of an enquiry committee along with famine initigation endeavours. In the Mizo Hills the administration never set-up any committee to investigate the matter. The government adopted a two-pronged strategy: one to reduce the menace of the rats and two, supply of foodgrains to the affected.

To encourage people to kill the rats, the government announced incentives for each tail of killed rats produced. By the end of 31 March 1913, 1,79,015 rat-tails had been produced to the government agencies. The government dispersed a sum of Rs. 1,532-6 *annas* to the villagers for the tails produced.[63]

The government had to distribute relief, and the total amount finally given out was Rs. 5,85,000. Since it was useless to hand out money to the people when there was no rice within the district to buy, this relief was given in the form of orders for rice at fixed price on shopkeepers at Sairang for Aizawl and Demagiri for Lungleh to which places rice was imported from outside the district.[64]

The government had already set-up trade marts and food godowns at various strategic points during the famine of 1880–1. These centres now became useful again. The government imported rice and paddy and stored them in these centres from where it was distributed to the people. Tipaimukh, Demagiri were earlier such centres. Now that the entire Mizo Hills were under British subjugation they could open such centres anywhere in Lushai country. Thus, new centres opened up at Sairang, Aizawl, Lungleh, Champai, North Vaiphai and so on. Each of the godowns had a head clerk as the in-charge who issued food grains to the relief seekers. The government entered into a contract with few food grain traders to supply rice and paddy to these godowns.

These contractors or *thekadars* as they were called were essentially big merchants of Cachar, mainly Marwaris, who would procure rice from the local market or transport them from neighbouring areas and send them by massive boats up to the Demagiri from where these were transported by bullock carts to the godowns at various stations. Although the suppliers received their payment from the government, it most interestingly did not distribute them to the needy free of cost. It was a matter of policy that rice would be sold from these godowns and if the villagers could not purchase them, these would be handed out as loan to be repaid in cash or kind in the next harvest after the famine was over. The loan would be claimed by the chief of a village or locality and would be handed over to him only. It was the responsibility of the chief to distribute the rice to the needy on the basis of government guidelines: the directive was that 'In distributing rice as a gratuity to indigent and infirm persons the allowance should be limited to the following scale per month: Adult of and above 15 years of age–29 seers; children from 6 to 14 years of age–12 seers; children below 6 years–5 seers.' Thus it was clear that rice would be given as gratuity, i.e. in lieu of service. In fact, the government referred to a specific case and issued directives 'It is stated that a certain old woman who used to gain a livelihood by selling firewood have since the free distribution of rice has ceased to do any work. Such persons who could do something towards their own support and had ceased to do so were given light, nominal work in Aizawl or Lungleh stations.'

The indications of another famine was apparent with the reports of flowering of bamboos in 1925–6. As per the predicted cycle the next bamboo flowering was scheduled for 1930–1. True to the cycle, the flowering had already begun, and with this the proliferation of rat population. The District Superintendent, N.E. Parry, immediately instituted a rat killing campaign which resulted in over half a million of them being killed.[65] The threat of renewed scarcity arising out of the flowering of bamboos persisted in the following year but fortunately did not materialize to the extent as it did in 1911–12. The famine of 1911 cost the government a sum of about £20,000, 'six times the total annual yield from the tax on house' whereas the famine of 1931 cost only about £1,000. This was because the government recovered most of the money spent on relief from the people through loan repayment.[66] In fact, this food for work scheme emerged to be a bone of contention. The Lushais although were hardy workers and believed in the theory of *Tlawmngaihna* or system of community obligation to public service did not like to be used by the British as coolies. In fact, in north-east India, the practice of forced labour among the tribals had been a source of many rebellions. The Kuki uprising (1917–19) was mainly sparked off by this practice, which the tribals resented.

By June 1912 the government had advanced Rs. 4.5 lakh. The cost of rice was about Rs. 65-10 *annas* per maund. Such food grains were issued to the needy but on the condition, as already mentioned, of repayment with interest. However, few people could do so due to the lack of money. The Lushai economy was basically a food gathering, subsistence economy level. Shifting cultivation and hunting were the dominant culture. Although the British introduced money, the monetization of the economy had hardly taken place. People did not have surplus production for sale. They still bartered their produce for certain scarce and necessary items. Hence there was no liquid money available with the tribals. The closure of the *duar* markets by the British stopped the exchanges between them and the plains people and as a result they did not have any money. All they were left with was agricultural products, which unlike other times could not be exchanged or sold in the *duar* markets. To top it all, 'tax' a new phenomenon, was introduced which the tribals were obligated to pay. Every household had to pay a levy ranging from Rs. 2 to 3 per year by which only the householder could procure the right of housing, from the government, to use forest resources and land for cultivation. There were also fees for grazing, fishing, garden produce, boat taxes, stall rents, registration fees and a number of fines under different heads.[67] Most of these were their birth-right before the British took over and taxes and fines were unheard of. With no purchasing power, question of repayment of loans taken during the famine period did not arise. Hence the state evolved the device of making the credit good by way of forced labour. Forced labourers were recruited in large numbers from villages to develop Aizawl, which the British were developing as the headquarters of the Lushai Hills. This labour force was requisitioned and transported to Aizawl where they were employed in road construction, construction of bungalows and offices for the British administrators and above all the construction of a water storage tank to supply water to Aizawl town as Aizawl was the most water scarce area in the hills. Such work for the whole day, away and uprooted from their respective villages for long periods was physically and mentally draining for the Mizos. They were not used to such long hours without a break. They generally would work hard and make merry within the community in the evening. But now it was not so. It caused great suffering to the Mizos. Besides the hardship caused by forced labour was the unfriendly attitude of the supervisors. The worst part was whether a person repaid his due in time or not, he would still be forced to go for labour in lieu of others who could not pay their dues in time.

Not only the construction works, the Lushais were made to carry the belongings of the officials and soldiers by headload and very often they were required to part with their eggs and chickens free of cost to the British officers. This caused great resentment among the Mizos. In fact, even the British authorities acknowledged that,

> I do not approve of the system of imposed labour, though it is absolutely necessary at present. It is about the only result of our presence that Lushai really object to strongly. I frequently say to them 'you force us to occupy your hills. We had no wish to come up here but you would raid on villages so we had to come and so now you have got to bear as much of the cost of occupation as possible, you cannot expect us spend money of the people of the plains on importing coolies, to do work that you are too lazy to do except under compulsion. These judgements admit to the Lushais who admit its justice.[68]

This summoning of compulsory labour was frequent since the inception of British rule. But during normal times people generally make excuses of *jhuming* work to be done and escape it. But during famines they could not do so. Moreover they were obliged to perform duties as they had borrowed a loan of rice which had to be repaid. Even if he had not taken any loan he had to work for the repayment of others. The only way to escape the hardship of famine and indignity of forced labour was to migrate out of the Lushai Hills. Villagers thus began to migrate by batches under the leadership of a chief or a dominant leader. The first batch moved from Serhman village under the leadership of Dokhuma Sailo to Tripura in 1911. The second batch recorded to have migrated was a group of 200 families under the leadership of Hrangvunga Sailo—chief of Bunghmun village. Actually this batch had shifted to this village from Serhmun village earlier. This batch now settled in the Jampui Hills of Tripura. It is recorded that these two batches migrated not due to the hardship of famine itself but the compulsion of working as forced labour.

Another curious fact of the famine-relief was that the administration issued an order to the effect that in each of the relief centres a list of names of people be displayed who were the recipient of relief and the amount of rice received. It is obvious that the administration wanted to publicize its generosity and the kind of help it was rendering to the people. The objective, though not mentioned, was obviously to make the Lushai aware that the Lushai were better off under the British administration rather than their own chiefs. Similar attempts were made by the British government in the rest of India where it perpetually tried to contrast the British rule with the eighteenth century, which it depicted as the dark age of Indian history. The Raj was successful in implanting this new image effectively.

The Government of India did not discriminate between the friendly and hostile chiefs in distributing famine relief. The idea was to win over the hostile chiefs as well through philanthrophy. The local officers, under the instruction of the higher authorities, provided work to the able bodied Lushais and with their help constructed good roads. The government, however, considered it undesirable that the Lushai refugees should settle down in the plains of Cachar and it was hoped that when the pinch would be over they should return to the place where they came. There was no denying the fact that without the large-scale liberal relief the famine could not have been tackled properly. It was believed by the Lushais that the English saved the Lushai population from an unprecedented calamity. What they overlooked was that the British came to their rescue only in two famines; they had survived many such famines in the past.

Changing Image of the Raj

The encounter between the Lushai tribals and the colonial power was an important event in the history of the Lushai people. It was significant not just in terms of an occurrence but also in its impact. A British administrator himself evaluated the situation in the following words:

> The advent of the British form of government and control to a time certainly paralysed the people. This was inevitable. Raiding excursions by Lushai had been countered abruptly by the permanent incursion of a power and a might far beyond the full comprehension of innate Lushai. All through the history of Lushai relations runs strong evidence of the respect and desire felt by the people for power. The Lushai was staggered, bewildered. The British occupation of Lushai marked the presence of a power, hitherto unforeseen and unimagined.[69]

The above statement is the usual boasting of a colonial administrator. Although his assessment of the Lushai being 'staggered and bewildered' was correct, it was not because of the presence of a mighty power, but rather at the rapid change of events in the life of the community. The Lushai were not conquered by British might. It is evident from the fact that the colonial power had been fighting against the tribals from 1826 and it was able to completely subdue them only by the first decade of the twentieth century. In fact, the tribals continued to harass the administration till that period even though the British had firmly established themselves at the heart of the Lushai Hills (Aizawl) in 1898 itself. The tribal could see through the British policy of sending emissaries after the failure of every military expedition. They often boasted to the neighbouring powers of Manipur and Tripura that the white power was actually paying a tribute to their chiefs by such acts.[70] But by the first decade of the twentieth century they began to submit themselves to the British, especially after the administration helped them out. It was not 'respect and desire felt by the people for power' but respect to a provider. The twin invasion of the British and the missionaries paralysed the traditional system. The people had no alternative but to fall back on the new system, which acted as the provider, even though at a price. It was not therefore military conquest but the moral conquest that settled the Raj in the tribal territory of the Lushais. Famine-mitigation contributed to this moral capitulation. The continued efforts of the colonial administration and the Church were able to relieve the distress of the famine-affected people to a considerable extent. Significantly, this produced a metamorphosis of the image of the Raj in the minds of the tribals.

The British first came into contact with the tribals of the Mizo Hills in 1826, when the latter raided the Sylhet plains. This began a long story of confrontation, warfare and punitive expeditions. When the British annexed the Cachar plains they also confronted the Mizos, who regularly raided the area. After the discovery of tea in Assam there was a rush to acquire the foothill lands for establishing tea gardens in the Cachar area as hell. As hill after hill were occupied southwards, it threatened the tribals, who feared that soon Europeans would invade and encroach on their homeland. Since then they had led a valiant fight against the white men, resisting every advance of the British towards their hills. In fact, they would often attack the plains, loot settlements, kidnap people and practice head-hunting on the British subjects. The tribals had always raided the plains for various purposes as shown in the previous section. But this time the nature of their raids was different. In addition to the earlier reasons it was also to register their protest and to scare the Europeans from invading the hills. The Europeans were objects of hatred for the Mizos. They were also amazed at the physical look of these newcomers. One was quoted as saying, 'These enemies [British] are different from other people who we have ever seen. They are white as goats. They clothe themselves from head to feet. They cover their feet with leather and we believe they will not able to climb the slopes of the hills.' The Europeans were also ridiculed for their white skins,

Bamboo plants with fruits

'Group of Lushai men and women before the government bungalow in Demagiri gathered to get free rice (1882).

Source: Willem von Schandel et al., *The Chittagong Hill Tracts: Living in a Borderland*, Bangkok: White Lotus, 2000. Riebeek, Emil, Die Hügelstamme von Chittagon: Ergebnisse einer Reise in Jahre 1882, Berlin: A. Asher & Co., 1885.

Bamboo flowers

Bamboo flowers

as 'half-cooked' people. But the same Europeans came across as kind and helpful people during the successive famine-related hardship. The colonial administration itself did not fail to observe that,

> A feeling of confidence and gratitude prevailed among the Western Clans. The fear and distrust of Europeans had diminished to such an extent the Assistant Commissioner of Hailakandi was invited in some cases to visit the Punjis. This is Noteworthy, for the Lushais are very zealous in excluding Europeans and are very superstitious in believing that Cholera always follows in their [white man] wake.[71]

The Church leaders who were active in distress mitigation during the 1911 famine also noticed the change and testified the transformation:

> In many ways we have been able alleviate the want and distress around us and gratitude of the poor people has been most pleasing to witness. Scores of men and women who had no food to eat have been enabled to go down to Demagiri to a fresh supply of food by the loan of a few pounds of rice apiece. Many others have been kept from want by being employed in building, road making, jungle cutting, gardening and other works about the compound. While not a few who have been unable to work have been assisted with gifts of rice. It has been a peculiar privilege to be living in Lushai Hills this year and thus be able to help the people in their hour of need. They have always looked upon us as their friends and at such times as this, the poor especially find our presence a source of comfort and strength for they feel that they come to us in their extremity and be sure of a helping hand.[72]

The same was true of the administration. The same report further states:

> Whatever feelings of resentment may have lingered in the hearts of some of these hill people against those who have occupied their country in order to prevent a repetition of their head-hunting raids upon the peaceful inhabitants of the plains, this famine must have surely dispelled it. For there are hundreds who would have starved to death this year but for kindly help rendered by the government in bringing up thousands of sacks of rice to supply their need.[73]

The above report enumerates the situation after the 1911 famine under British rule. Since then, more famines had stalked the hills of Mizoram. The relief measures provided by the administration had a profound effect on the overall image of the Raj in the minds of the people, who began to look up to it as a kind and merciful system manned by white-skinned Europeans. The administration was looked up as paternalistic, and in a matter of two decades the white men were now addressed as *Saab-Pa* (white-father), *Mirang Bawipa*, *Mikang Topa* or *Mirang Topa* or *Mirang Lalpa*, meaning White Master, White Father, Lord of the Lushai Slaves, or even the White Lord. The Lushais began to rechristen the white officers and missionaries with Mizo names. One British officer, T.H. Lewin, was so popular among the Mizos that the villagers called him as Thangliana—a Mizo name. D.E. Jones the Welsh missionary who pioneered the presbyterian church in the Mizo Hills was called Zosaphuia and J.M. Llyod, late missionary of the presbyterian mission, Zohmangaihi Pa. Similarly a number of British officers were given fond names like Zo-saap, Zosaaphuia, Zohmangaihi Pa, Zo-paa and so on. What is noteworthy is that all such names had paternal implications. This is what the British always tried to achieve: to conquer the Mizos morally so that they were ethically bound to them. This was what they achieved. They wanted to project the British as a paternal figures who protected the subjects, secured them from enemies, provided them succour from calamities and the colonial administration as a paternal system. From the results it was obvious that the Raj had been successful in manufacturing the desired consent to their rule and implanting that image.

The impact of famine relief on the image of the Raj can be better understood when contrasted with the indifference of the Indian state when a similar famine occurred in 1957 after the British left. People constantly referred to the benevolence and kindness of the British as against the indifference of the post-colonial Indian government, and recalled how they were better off during British rule; when their distresses were cared for when natural disasters took place. It may be repeated that the insurgency in Mizoram started only after the 1957 famine, and that the Mizo National Front, which led the secessionist movement was originally a voluntary organization called the Mizo National Famine Front, formed to coordinate famine relief fronts. The lamentation of a Mizo poet after the Indian government launched repression against the Mizos, poignantly expresses the peoples' perception of the Raj:[74]

> I dare not contemplate this grief of our land.
> Departed are our civilised white skinned masters.
> Oh, God who succour the poor,
> I pray thee
> Set the tottering land on its feet once again.
>
> (translated from the original Mizo)

Critical appraisal of the political economy of relief however, shows a different picture. Bamboo flowering and the consequent famine were part of Mizo history. They had experienced it several times in the past. No doubt it took a heavy toll on their population but they had also evolved mechanism to cope with it. But the colonial state destroyed it and made them entirely dependent on the administration. The most important part of this traditional coping mechanism was migration to the non-famine area or to the plains. The British stopped both through legislation. Even the migratory routes were closed and movement towards a better habitat, which was natural to a nomadic tribe, was halted. Tribes and sub-tribes were forced to settle permanently in a particular village and were not allowed to abandon it, which was their tradition. The slash and burn cultivation could not be practised in the same land as forest cutting and burning would reduce the fertility of soil and the tribes would be forced to move to a new site. Traditionally, with the shift in farming land, the habitat also shifted. The same was true of water, which was very scarce in the hills. Once a water spring dried with the onset of the dry season, the village had to be shifted. But now that the Lushais settled permanently in one village they could not shift to a new site either for cultivation or for water. They had to change sites within a particular range centred round their village. In fact, they could not change many sites, as with the introduction of the concept of private property, villages had their sites demarcated leaving no territories free for occupation or cultivation. The diminishing fertility of land and scarcity made the tribe depend on the state for food and water. The administration had to arrange for food and water supplied to the villagers, though for a price. Thus the state had managed to make the Lushai tribe dependent on the administration for their essential consumer items even during normal times. In famine times the situation was worse. The diminished productivity of land was compounded by the destruction of food crops by the rodents. The traditional extra-cropping in famine year was also given up. The tribe, which would earlier take precaution months ahead according to their calculated year of bamboo flowering, failed to do so as their prediction did not match the records of the administration. Even a basic precaution like large-scale killing of rats was to be induced by the administration. In fact, the 1911 famine showed that the administration offered pecuniary benefits to the tribals to kill rats. This was just 30 years after the last famine of 1881. Similarly migration to the plains was abruptly stopped by the administration. The administration introduced the concept of 'boundary' between habitats. Migration was not allowed even during crisis like famines. The famine as seen above was a time of hectic commercial activity between the tribals and plainsmen. But with the closure of these migrations, the exchanges—both commercial and social—also stopped. The 1881 famine showed that the Lushais had enough money through exchanges of their products to buy their own food from the government storehouses. The government itself reported it. But in the 1911 and 1931 famines no such exchanges were noticed. The tribals instead borrowed food from government storehouses and worked as labourers to repay his loan. They stopped eating forest products such as jungle *yams* or jungle rats as contact with the west through administration and Christian missionaries made them abandon certain food habits. The destitute families also could not offer themselves as *inpuisang bawis* (voluntary slavery) to the chiefs as after the *bawi* controversy of 1911–12, which will be seen in the next chapter, the practice was stopped. Thus the entire traditional social security system of the Lushai tribe to cope and combat a formidable natural calamity was crushed with the advent of the colonial state. Their support system was paralysed and made them totally dependent on the colonial administration. The tribals were still grateful to the administration for mitigating the disaster for them. They never realized that even though it cost a number of lives, after every *mautam* or *thingtam*, the Lushai would still be sovereign and free to tackle the next calamity the way it wanted; but after 1881 they were not free anymore to even fight their own battles.

Even at a more material level, the balance sheet was not in favour of the Lushais. A simple quantitative analysis would be sufficient. According to the *Administrative Report* of 1913–14 of Lushai Hills during the famine of 1911–12, the government had expended a sum of Rs. 5,39,927-11-0 towards the cost of importing foodstuff for the Lushai Hills. Since most of the relief was given as loan, to be repaid in cash or kind, the government received back the majority of the loans of rice in the next harvest. There was a policy behind this 'loan' too. The government resorted to a loan instead of free relief, since the recipient of the loan would remain legally obliged to the administration until the full repayment of the loan. In fact, most of the tribals were unaware that it was a loan which had to be refunded. Hence, after the famine, when they learnt of their obligation they were devastated. They had hardly produced any surplus to repay old loans. They had to render manual labour to pay this obligation. The rest had to serve the administration as forced labour against the loan. In a similar instance, when a famine occurred in Zemi Naga areas of north Cachar in the 1940s, the tribals refused to take any relief as a loan from the government fearing return obligations in cash or compulsory labour, which they detested.[75] The government built a host of houses and office buildings for its officers through this labour force. It had even constructed a huge water reservoir, which could supply water to the whole of Aizawl town with this labour. This labour was virtually free. In other words, the government had a network like trade marts, road and path constructed with half the money it spent on relief. So much for Paternalism.

POLITICS OF PATERNALISM

The colonial authorities claimed to be upholding the principles of Paternalism. In fact, throughout India the British often claimed that they endeavoured to develop a paternal system of administration for the backward people of the subcontinent—a government, which would work as a deliverer, a protector and a liberator. Here they were not really talking about the nineteenth-century English philosophy of Paternalism. They meant it literally—of being a paternal figure to the subjects, which as a notion was more medieval in nature than befitting a modern colonial power. But the British used it as a rhetoric to counter the adverse notions that people had about the Raj in the early nineteenth century. British made a similar claim of paternalism in north-east India. In their very first encounter with north-east India they put forward this theory. After the Treaty of Yandabo (1826) which concluded the Anglo–Burmese War, the British distributed leaflets in Assam stating that they had entered Assam only to protect its people from the atrocities of the Burmese invaders and that they had no intention of staying on in the region.[76] Yet, soon the paternal protection gave way to the rosy prospect of adding another province to their growing empire in India and by 1835 they had usurped the authority of the native king and assumed power themselves.

Like most of the other ideas of the period, such as Utilitarianism, Evangelism, Paternalism too emanated from English liberalism. The Paternalists, which included thinkers like Thomas Munroe, Thomas Metcalf, Malcom and Elphinstone opposed the Utilitarian attempt to uproot the traditional structure of Indian society and anglicize them. For them the function of the government was simply one of paternal protection. In fact, for Munroe, an ardent disciple of Edmund Burke, the first task of British power in India was paternal protection and little more. Indeed, the common aim of the Paternalist school was to conserve the original institutions of Indian society, rather than to construct society anew. This vision of a benevolent paternalism was founded on the concept of unchanging village republics and they never contemplated a system of Direct Rule that would remould India in the image of the West.[77] In, fact Munroe candidly stated, 'The Englishman are as great fanatics in politics as the Muhammedans in religion. They suppose that no country can be saved without English institutions. The natives of this country (India) have enough of their own to answer every useful object of internal administration and if we maintain and protect them the country will in a very few months settle itself.'[78] Yet despite the claim to Paternalism this is what they did not allow to happen.

The Agent to the Governor-General in Assam, David Scott, the architect of the expansion of the British Empire in north-east India was an avowed Paternalist ideologue. The men who was responsible for the annexation of Assam, credited for the conquest of 'the hitherto uncompromising independent tribes of Garo, Khasi and the Singphos' ironically was also described a Paternalist.[79] Like the father of Paternalist philosophy—Munroe, Malcom, Metcalf and Elphinstone, he was also Conservative and was opposed to the attempts to assimilate and anglicize everything traditionally Indian.[80] He even went to extent of re-establishment of the institution of slavery in the name of restitution of tradition during the early period of colonial rule in Assam. At the same time he invited Christian missionaries to 'reclaim' the 'savage' tribes to civilization in order to render them fit to be colonial subjects. Such contradictions were therefore quite common among the English imperialists. It was evident that the real issue was the interest of the Raj and everything else, even ideologies were subordinate to the objective of imperialism. Thus when the British administration in Cachar came forward to mitigate a natural calamity in the Mizo Hills, the avowed claim was philanthropy and the paternal face of the white men. They had not yet subjugated the tribes but were greatly harassed by them. Despite their repeated attempts they failed to subdue them. They felt that the demonstration of the paternal patronization during a grave crisis would prompt the tribal to reciprocrate with friendliness.

The British encountered the ferocity of the frontier tribes like the Mizos from the 1820s. The resulting encounters lasted well over five decades and even by the 1880s when the famine occurred in the hills the British were nowhere near subjugating the tribes. The growing tea industry was facing closure due to the tribal raids. The tea-planters as well as public opinion in England was severely critical of British power in north-east India for failing to curb the atrocities of a small band of tribal raiders. Obviously, the British administrators in the frontier did not entertain any paternalistic attitude towards these marauding tribesmen.

There was a slight shift in the paternalism of this period from the earlier ones. The early paternalism of the Raj followed strict non-intervention in tribal affairs. But the perpetual raids compelled them to go for outright coercion, which also failed. The *mautam* gave the British, the opportunity to show case their paternalist image. Right from the District Administrator to the Lieutenant-Governor all had foreseen the potency of this weapon and hence did not oppose the huge investment in relieving the distress of an erstwhile enemy. The tea-planters however were opposed to the idea but once convinced of the covert agenda of the administration, they too participated in famine relief even if they would have liked to use this opportunity to get back at the hated head-hunters. But if they did so, the high ethical standards which the Victorian liberals had laid claim to, would not be justified. The cumulative result was the same. What 50 years of warfare could not achieve, one year of famine relief did. Within the next few years the colonial power was firmly settled in the heart of Mizo Hills in Aizawl.

Immediately after the famine was over, when the younger generation of the Mizos was eager to go back to their raids and plunder of the British territory, the chiefs opposed it. They were against harassing the white people who provided them protection during the time of crisis. This paternalist image of the white people was gradually reinforced with the establishment of the British power in the hills. After the first one, the British witnessed three more and with each of them their image was consolidated. The Mizo, who consider every non-Mizo as *vai-chia* (bad foreigners) developed affectionate terms for their white masters. In fact, the English were never called *vais* (aliens), even though they were one. Now the Mizos added the appellation *pa* (father/fatherly) to the Englishman. At the most, an outsider would be called *kapu* (sir), but Englishmen were given names like *saabpa* (fatherly white men), *mirang bawi pa, mikang to pa, mirang to pa*, or *mirang lalpa* meaning White Father, White Lord, as mentioned, earlier in this chapter. From the practice of Paternalist politics to the emergence as an actual *paternal figure*, the British colonialists in the Mizo Hills had indeed had a curious metamorphosis.

NOTES

1. John Butler, 'Rough Notes on Angami Nagas', *Journal of Asiatic Society of Bengal,* vol. 44, no. 4, 1875, pp. 307–27. Cited in Asoso Yonou, *The Rising Nagas: A Historical and Political Study*, Delhi: Vivek, 1974, p. 72.
2. B.C. Allen quoted in Hokishe Sema, *Emergence of Nagaland*, Delhi: Vikas, 1986, p. 15.
3. A.G. McCall, *Lushai Chrysalis*, London, 1949; rpt., Kolkatta: Firma KLM, 1977, p. 65.
4. Edward Gait, *A History of Assam*, Calcutta: Thacker, Spink & Co., 1905; rpt. 1984.
5. *General Administration Report*, 1891–2, CB-1, 6-9, Aizawl: Mizoram State Archive.
6. Details given in the last chapter.
7. *Bengal Judicial Proceedings*, 1850, 27 February, nos. 36–7, in H.K. Barpujari, *The Comprehensive History of Assam*, vol. IV, Gauhati: Publication Board, 1992, pp. 150–1.
8. *Bengal Judicial Proceedings*, 1859, October, no. 7, in Barpujari, ibid., p. 177.
9. *Foreign Political Proceedings*, 1869, nos. 72–105, in Barpujari, op. cit., p. 178.
10. *Foreign Political Proceedings A*, December, no. 245; *Bengal Judicial Proceedings*, 1869, August, nos. 222–3, in Barpujari, op. cit., pp. 179–80.
11. Ibid.
12. *Foreign Political Proceedings A*, September, no. 249, Edgar, 5 June 1872.
13. Ibid., June, no. 128, Lewin, 26 March 1872.
14. Ibid., September, no. 269, Atchinson, 4 September 1872.
15. D.R. Lyall, Commissioner of the Chittagong Division, Bengal Secretariat, Pol. A, June 1891, nos. 1-139, File L/20 1889.

16. Ibid.
17. C.J. Lyall to the Government of India, Revenue and Agriculture Department, 26 July 1882 in Foreign Political Proceedings, no. 89.
18. Col. E.B. Elly, *Military Report on the Chin-Lushai Country*, Simla: Government of India, 1893, pp.14–15. Also Suhas Chatterjee, *Mizoram Under British Rule*, Delhi: Mittal, 1985, p. 96.
19. McCabe during one of his expeditions to the Lushai Hills wrote, 'Exposure and Starvation are our strongest allies and with their assistance I believe that the Lushais will very shortly be craving for peace.' Cited in McCall, op. cit., pp. 59–60.
20. 'Report of the Scarcity of Food in the Lushai Country' from J. Knox Wight to the Chief Commissioner of Assam, Silchar, 9 June 1882, in ibid.
21. Ibid.
22. Ibid.
23. Ibid.
24. Ibid.
25. Suhas Chatterjee, *A History of the Mizo Economy*, vol. I, Jaipur: Printwell, 1999, pp. 192–3.
26. Ibid.
27. Ibid., p. 152.
28. McCall, op. cit., p. 41.
29. Ibid., p. 48.
30. Chief Commissioner of Assam to Rivers Thompson, Foreign Political A, January 1882, pp. 795–815, dated 30 November 1881, NAI.
31. Chatterjee, as in note 25, pp. 152–3.
32. Chief Commissioner to Deputy Commissioner, in Foreign Political A, January 1882, 795–815, dated 30 November 1881, NAI.
33. J. Knox Wright to the Chief Commissioner of Assam, Silchar, 9 June 1882, no. 860 J, *Proceeding of the Foreign Department*, vol. I, Pol. A, August, no. 88-91.
34. Chatterjee, as in note 25, pp. 152–3.
35. Ibid., p. 152.
36. Ibid.
37. H. Brown, *The Lushais, 1888–89*, Shillong: Govt. of Assam, 1890; rpt., Aizawl: Govt. of Mizoram, 1978, p. 68.
38. Chatterjee as in note 25, pp. 152–3.
39. M.O. Boyd, ICS, Deputy Commissioner, Cachar, Silchar, 14 March 1881.
40. Ibid.
41. Ibid.

42. Ibid.
43. Ibid.
44. Ibid.
45. Ibid.
46. *The Pioneer*, London, 11 June 1870.
47. *The Observer*, 25 February 1871.
48. *The Pioneer*, London, 22 February 1872.
49. Ibid.
50. Chatterjee, as in note 25, p. 155.
51. Ibid.
52. Ibid.
53. Ibid.
54. Ibid.
55. Ibid.
56. Ibid.
57. *General Administration Report, Government of Assam, 1874–5*, p. 41.
58. Ibid.
59. John Knox Wright to Secretary, Chief Commissioner of Assam, 1.F (External) A, October 1884, no. 380.
60. Ibid.
61. Robert Reid, *History of the Frontier Areas Bordering on Assam from 1883–1942*, Shillong: Govt. of Assam, 1942; rpt., Guwahati: Spectrum, 1997, p. 45.
62. F.C. Hennikker, *Maotam in Lusai: Notes Compiled in July 1912*, Maidstone, 1930, Hennikker Papers, Box no. 10, Cambridge: Centre for South Asian Studies, University of Cambridge, p. 4.
63. C. Rokhuma, *What is the Anti-Famine Campaign Organisation Doing*, Aizawl: Author, 1988, p. 102.
64. Reid, op. cit., p. 45.
65. Ibid., p. 47.
66. McCall, op. cit., p. 169.
67. Ibid., pp. 228–9.
68. *General Administration Report*, 1891–2, CB-1, 6–9, Mizoram State Archive.
69. McCall, op. cit., pp. 196–7.
70. Ibid., pp. 34-64.
71. Knox Wright to the Chief Commissioner of Assam, Silchar, letter no. 860J, 9 June 1882, *Proceedings of the Foreign Department*, vol. 1, Pol. A, August 1882, nos. 88-91.

72. *Annual Report of the Baptist Mission Society*, South Lushai Hills, Assam, 1912.
73. Ibid.
74. Recorded in V. Venkata Rao, H. Thansanga and N. Hazarika, *Century of Government and Politics in North East India*, vol. III:, *Mizoram*, Delhi: S. Chand & Co., 1987, p. 267.
75. Ursula Graham Bower, *Naga Path*, London: Readers Union, 1952, pp. 159–60.
76. Amalendu Guha, *Planter Raj to Swaraj: Freedom Struggle and Electoral Politics in Assam 1826–1947*, Delhi: ICHR, 1977, p. 1.
77. Eric Stokes, *The English Utilitarians and India*, Delhi: Oxford University Press, 1989, p. 18.
78. Cited in ibid., p. 19.
79. Nirode K. Barooah, *David Scott in North East India: A Study in Paternalism, 1802–31*, Delhi: Munshiram Manoharlal, 1977, p. 1.
80. Ibid., p. 231.

CHAPTER 4

Famine as a Site for Politics of Humanitarianism

As 'empire' became an established fact, there was an increasing inclination in England to transform the colonies in its own image. The endeavour to complement political conquest with cultural colonization also started. Often this was done through 'humanitarian methods'. Fresh from the Renaissance, the Europeans had evolved new parameters of humanism on the basis of which principles of humanitarianism was constructed. In England one of the pioneering attempts came from the Clapham Sect. The sect started with two primary objectives—abolition of the slave trade in the colonies and opening them to Christian missionary enterprise. They had their first major victory when permission was granted for missionary activities in India in the charter of 1813. The Bible thus followed the flag. The introduction of Christianity reinforced the process of conquest. In the hills of Mizoram, the physical as well as cultural conquest through evangelization went on side by side.

The missionary project was however was not as autonomous an enterprise as it is often presented to be by the ecclesiastical and a section of secular historical literature. There was a crucial connection between the two, which has been the subject of an acrimonious debate between secular and ecclesiastical historians. The organic correlation between the empire and Christian missions has to be traced to the beginning of the evangelical movement in England. An ideal instance of collaboration between missionaries and colonialism was strongest in case of Spain and Portugal where the colonial state not only sent out missionaries but also looked after their protection and sustenance. In contrast, the English State used the missionaries to legitimize colonial rule. As a result the missionary endeavour did not turn political. Even though the missionary had not served the colonial power directly, it was only too glad to accept protection and development from it. Often they tried to justifiy colonial rule as 'Divine Command'.[1] The concern for the spiritual progress of the colonized, euphemized as 'natives' and the promotion of the 'ideas of advantages of civilization' was taken up in the Congo Conference of European States held in Berlin in 1885. The Conference identified 'civilization' with Christianity and 'savagery, barbarism and paganism' with the natives and discussed the necessity of spreading the gospel amongst them.[2]

The East India Company's attitude towards the missionaries changed according to political expediency. Even though it assumed political power in Bengal after the Battle of Plassey, it was wary of civil rebellion. Any Christian missionary work was disfavoured because it was feared that such attempt would incite intrigues and disturb the political stability.[3] Back in Britain however, the rising middle class was bursting with ideas of political and social movements. These middle classes wanted to experiment with these ideas and what better laboratory could be there than the virgin fields of the colonies. The fountainhead of most of all these ideas was the Liberalist thought in England. 'The material and intellectual elements which composed English liberalism in India was threefold: Free trade was its solid foundation, evangelism provided its programme of social reform, its force of character and its missionary zeal. Philosophic radicalism gave it an intellectual basis and supplied it with the science of political economy, law and government.'[4] The transformation of the English in India from suppliant merchants to a ruling class was based on a sense of racial superiority, which was a natural consequence of this graduation. With this there was a coeval improvement in its moral tone—the emergence of a new ethic in this new society which desired the transformation of colonized 'native' societies to 'higher' standards—the parameter of 'higher' being those set by the European colonizer.

Originating in the form of Methodism, the new outlook took the shape of evangelical revival in the late eighteenth century among the upper middle classes of English society. It was this revival and the activism of the English middle class that resulted in the formation of the Clapham Sect by Charles Grant, John Shore with Zachary Macaulay, Henry Thornton and John Venn.[5] The two great objectives of the Clapham Sect were to secure the abolition of slave trade in the colonies and opening up of the colony of India for missionary enterprise. Sensing the opposition of the East India Company to open India for missionary enterprise, Charles Grant produced a treatise entitled 'Observation on the State of Society among the Asiatic Subjects of Great Britain, Particularly with respect to Morals and Means of Improving It' in 1797 and placed it before the Court of Directors of which he himself was an influential member. Grant's objective was to counter the Anti-Church ideas of Thomas Paine and revolutionary principles propagated by France. His ideas were with an eye to the preservation of empire. 'Christianity of the English sort might keep Indians passive just as it induced contentment in the English lower order.'[6] By 'English Lower Order' Grant was referring to the tension among the English working class which was neutralized by the use of religion through by evangelical revivalism in the late eighteenth century. Grant's plea for Christian mission in India was motivated more by political conservatism than any radical ideology. He was one with Bishop Horne's view that *English Christianity* would inculcate, 'in superiors it would be equity and moderation, courtesy and affability, benignity and condescension; in inferiors, sincerity and fidelity, respect and diligence. In Prince's justice, gentleness and solicitude for the welfare of the subjects; in subject's loyalty, submission, obedience, quietness, peace, patience and cheerfulness'.[7] Grant was also convinced the conversion to Christianity would ensure permanence of the Raj in India. More astonishing was his belief that the Raj was providentially ordained. At the same time he was aware that the main objective of the East India Company was commerce. 'In every progressive step of this work we shall also serve the original design with which we visited India, that design still so important to this country—the extension of our commerce.'[8] Grant was apparently not unconcerned about the Indian 'natives' too. The poverty of the Indian people and their 'unformed taste' were considered to be the main hindrances, which limited the penetration of British manufacturers into India. Grant was optimistic that Christianity and education—the two 'noblest species of conquest', could remove these obstacles. Although the British government's response to the missionaries was unpredictable, there were others who favoured missionary intervention. Charles Simpson at Cambridge provided spiritual leadership. Edward Parry, Chairman and Charles Grant, Vice-Chairman of the Court of Directors in their letter to the President of the Board of Control argued that the Christianity could be the bond between India and England. 'If . . . they embrace our religion, they would have a new cause of attachment to us . . . which would give us better assurance of their fidelity.'[9] With such high officials and influential members of Parliament behind the cause and the unfailing support of the Evangelical party, William Wilberforce, a member of at least 70 philanthropic organizations in England, moved the English Parliament to secure the opening of India to missionary enterprises. Wilberforce collected 837 petitions from different Missions in support of his cause. In course of the debate, Wilberforce argued that *the sole justification and strengthening of British control over India lay in conversion of the Indian people.* He even appealed to his countrymen to transplant their principles, laws, institutions, manners and above all religion and morals on Indian soil.[10] The support to the cause came from another unexpected quarters—the Free Trade merchants. They contended that Christianity would change the habit and manner of people thereby increasing the demand for British manufactured goods. Clause XXXII of the Charter 1813 allowed the propagation of Christianity and unrestricted entry of missionaries for the purpose to India thenceforth. The Charter of 1813 was a triumph of colonial-evangelical collaboration.

One of the earliest missionaries in the Mizo Hills, R.J. Lorrain alluded to the Raj-missionary connection in his autobiography,

Five Years in Unknown Jungle for God and Empire.[11] But not all missionaries in the Mizo Hills were for Empire or supportive of the administration. This was not only because missionaries in these remote parts of India were not all British but also there were clashes of ideas on the issue of the degree of conversion. For the missionaries conversion involved not just Christianization but also 'reclaiming' the people from savagery to 'civilization'. To them civilization was identical with Westernization[12] which in turn was equalized with modernity. The association of Christianity with the West was really curious, as it was originally an Eastern religion. It came to the colonies through the West due to which it was branded as a Western religion. Even in Europe Christianity was not a single homogeneous category and there were a number of division and denominations of Christianity. This was the reason that proponents of Christianization in the colonies emphasized that in the English colonies it had to be English Christianity.[13] English Christianity, however, involved not just conversion to a new faith but complete transformation of the natives—from attire to attitude, food to festivals.[14] On this issue the colonial administration in India had quite different ideas. The two agencies therefore often found each other in confrontation. In this sense they were autonomous, but mutually interdependent for sustenance. They reinforced each other. While the colonial state conquered in tribals, the missionaries rendered them conquerable. The administration governed them and the missionaries made them governable. One conquered them politically; the other consolidated it by conquering them morally and culturally. The former looked after peace, law, and order while the other established new social and cultural institutions. The conquerors annexed them to the Empire while the missionaries made them English. If the colonial state had made them dependent on the British by disintegrating their traditional political, economic, social, and cultural support structure, missionaries consolidated that dependence. The rat-famines resulting from periodic bamboo flowering proved to be the perfect catalyst in this collaborative project of subjugation.

Amidst The Head-hunters

George Nathaniel Curzon had insisted that the empire was not the result of natural vanity or outgrowth of territorial cupidity. On the contrary it was a noble commitment. There had been intense human suffering on the part of colonists in the undertaking of colonizing lands which were infested with cholera, malaria and enteric fever.[15] The missionaries too felt the same. For Reverend Harry Inglish, a pioneering missionary, evangelical work meant 'travel through rough tracts of scrubs, jungle, heavy forests, across gullies and river-beds, up and down the slopes of hill carrying with him a few medical comfort to *semi-savage people, jungle tribes, aborigines, poor, primitive souls and bodies yet higher than animals*'.[16] All these hardships only 'lighten the darkness of these miserable beings and give them spiritual and physical aid'.[17] The Christian missionaries who visited the Mizo Hills had decided that they would work amongst the savage tribes. A people who hunt for human skulls had to be savage. Reaching thses hills involved *travel through rough tracts of scrubs, jungle, heavy forests, across gullies and river-beds, up and down the slopes of hills.*[18]

The colonial ruler had invited and welcomed the missionaries into the North-Eastern Frontier not so much for evangical purposes as it was with political and security reasons. Since the Diwani of Bengal (1765) the colonial authorities came across these tribe who were virtually naked, lived by hunting and food gathering and constantly harassed the plainsmen in the border area through raids, plunders, kidnapping and head-hunting. They first came across the Garos and the Kuki–Lushais on the Bengal frontier and then the Nagas on the Assam frontier. As the British advanced into the interiors, more so after the successful beginning of tea plantation, the harassment of the border people as well as the white tea planters increased. As the ruler of the area it was the duty of the British to protect its subjects from these bloody perpetrations. But number of violent expeditions into the hills failed to yield any substantial results. Then it was decided to invite the Christian missionaries to work among these tribes. The idea was that, once the tribes, who belonged to an unorganized and un-institutionalized animist faith, were converted to Christianity; they would not view the white men as hostile foreigners and would then succumb to peaceful co-existence. It was David Scott, a civil servant in Assam (1804-31) who first sought the assistance of the Christian missionaries in 'civilizing' the Garos to prevent their outrages.[19] Jenkins also confirmed, 'To put an end to their outrages, there could be no other means than a reformation of their feelings and habits through Christian religion.'[20] Although the ostensible catalyst of the Christian missionary works in the hill areas of the north-east was a civilizing mission, the real push was the 'colonial conquest' of the hill tribes. This is evident from the fact that even though the Charter Act of 1813 of the British Parliament removed the restrictions on missionary activities in the colonies, the East India Company was opposed to such activities. It feared that this form of cultural intervention would disturb the peace in the colonies and eventually jeopardize the Company's commerical interests. Yet the same Company's government decided to invite and allow the missionaries to work among the hill tribes when its military might failed to vanquish the tribes which posed an incessant threat to its frontiers. The hypothesis was that the missionaries through evangelizing would 'civilize' the 'savage tribes' thereby 'tame' these 'unruly' elements. Thus even before the missionaries had any idea of the tribes, the colonialist had formulated that the tribes were 'savage' and 'uncivilized' and the Western Christian influence which was 'civilized' would be able to tame and conquer them culturally.

The missionaries who came to the region therefore already had a pre-formed opinion that they were to 'civilize' the 'savages' and the way to do it was through evangelization. In other words, nakedness, archaic methods of food gathering, belief in animistic faith, kidnapping, raiding and 'head-hunting' were already defined as features of savagery and uncivilization. Civilization was equalized with Western living and since the West professed the Christian faith, it was the only religion of the civilized. The missionaries came with this mindset of superior civilized people encountering a lowborn savage and evangelization was a destined white man's burden, which they had to carry. In their acts and attitude towards the tribal, they reflected this attitude of superiority and pity at the degradation of hill people. The early missionary texts pertaining to the tribes are full of pejorative and uncharitable references. Even before they reached the Lushai Hills they were convinced,

The inhabitants were regarded by the few Europeans then residing in Bengal, as the fiercest and most barbarous of all the tribes within the province, notorious for their head-hunting expeditions to neighbouring plains. The object of these raids was to obtain human skulls with the object to adorn the graves of their ancestors, the belief prevailing that the spirits of the slain would become the slaves of their ancestors in the spirit world.[21]

One of the first missionaries in Lushai Hills, D.E. Jones of the Welsh Calvinistic Church described the land as: 'It is a country well-fitted for a Christian missionary with so many languages, cultures and barbarism living side by side. Some of its people are still in the stone age.'[22] The missionaries described their life amidst the tribals as 'living in horrors of savage warfare'. In fact, they were constantly haunted by fear of head-hunting. The Lushais were 'pagans' and evolved their own religious practices. They used to worship *Pathian* the primeval spirit and *Ramhuai*, the evil spirit. To the missionaries tribal religion was nothing but 'demon worship'. 'All these hill people were demon worshippers; but each tribe has its own demons and its own ceremonies, preserved in pristine purity or largely modified by their environment.'[23] As far as the religious faith of the tribes were concerned, it was repeatedly emphasized that their religion 'was a crude form of demonology. Its main principle consists in the endeavour to propitiate evil spirits by the offering of sacrifices. A vague belief exists in a supreme being.'[24] The tribals were often associated with liquor and drunkenness, which the missionaries detested. The Mizos were said to be 'lazy, cruel, superstitious and very prone to drunkenness'.[25]

Another characteristic, which the missionaries associated with tribalism, was their marriage institution. They often frowned upon the sexual openness of the tribals but what they strongly condemned beside polygamy was the ease with which tribal men abandoned their wives. In other words, the absence of divorce laws and the weak institution of the marriage itself were associated with tribal life.

The worst feature in the manner of the people and one likely to be a serious obstacle to the missionary is the laxity of their marriage, indeed divorce is so frequent that their unions can hardly be honoured with the name of marriage.[26]

Similarly, the missionaries also detested the tribal culture of breaking out into community singing, dancing and feasting', where animals were sacrificed and spirit worship practised, accompanied by drunkenness. The missionary in their evangelization effort tried to discourage such performances. In fact, the missionaries reported that the opposition and hostility to Christianity and reversion back to 'heathenism' by the tribals was exemplified by the elaborate dance and song performances by the community. While tribalism was condemned as 'savagery' the missionaries in their writings constantly referred the fields of their work as *beyond the boundary of civilization, unknown jungle, outside the civilized world.* But they rationalized their presence in the area as 'ordained' to bring the words of the Saviour to the tribal.

Wherever we go and the more we hear of the customs, habits and lives of the people, the more we are convinced of their need of the Saviour. Deeper is the shadow of this country's sin than the dark hue of surrounding mountains under and approaching storm, brighter are our expectations and hopes than the coming down of the eastern sky, for we want the coming of the great light of the world who shall pour forth His eternal light on these people who sit in darkness and in the valley of the shadow of death. We wait for the coming of Him who bringeth life to those who are perishing for want of truth.[27]

The New Head-hunters

Having defined the Lushais as 'savages' the missionaries set out to conquer them for 'God's Kingdom'. They searched for the first convert and when in 1899, two Lushais were baptized to Christianity they registered their first success. Since then, head counts of Christians among the Lushais began to increase. The Mizos in the later years used to describe the census data collectors as head-hunters because they were always counting heads. The missionaries were thus the new head-hunters who counted the number of converts. Annual reports were diligently maintained to enumerate the growth of the Church and report every new convert. These were moments to rejoice.

The first missionary to visit the Mizo Hills was Reverend William Williams of the Welsh Calvinistic Church in 1891. W. Williams was a young Presbyterian missionary in the Khasi and Jaintia Hills which lie several hundred miles north of the Lushai Hills. Among the Khasis there was already a very sizeable Christian following. He wanted to establish such a Church among the Lushai and decided to travel south to visit the country. In early 1891, four months after the Mizos killed Captain H. Brown and a number of others, W. Williams took an arduous journey to the Lushai Hills and stayed at Aizawl for a month. Williams found the Lushai Hills to be a potential area for missionary activities.[28] It was due to his persuasion that the Welsh Presbyterian Assembly decided adopting Lushai Hills as a mission field. But the sudden death of Williams put the proposal under a shadow. About two years later, J.H. Lorrain and F.W. Savidge reached the Lushai Hills as missionaries sponsored by R. Arthington of Arthington Aborigine Mission, on 11 January 1894. Savidge and Lorraine came from Sadia (Assam) to Aizawl in 1891 with the avowed object of spreading the gospel among an aboriginal people. They failed to do anything in this respect for the first three years as the Lushai Hills was still under a rebellion against the British. The Lushais were very suspicious of all white men. The government also did not look upon their activities with favour as it felt any interference in tribal culture would incite further resentment against the British. The Lushai administration was already finding it very difficult to govern the tribals. While the administration was uncooperative, there was not much headway in their proselytization efforts too. The first thing that put the two missionaries on disadvantage on their arrival was their ignorance. To overcome that they learnt the Mizo language, started writing it with the Roman Script, they taught a number of Mizos reading and writing, translated Gospels of Luke and the Bible in the Mizo language and a number of other books based on the Bible for use in Sunday School. They also wrote a book, *Grammar and Dictionary of Lushai Language* and also worked for the propagation of Christianity. They preached mainly in the villages of north Mizoram. Savidge and Lorraine did their best in popularizing education among the Lushais. Their Mission was the centre of spreading education among Lushai boys and girls. The missionaries went from door to door to apprise the Lushais of the necessity of education. At first their centre was at Aizawl, but later they changed their site to another village belonging to chief Saipuia. They were very popular among the Lushais and their services greatly helped the spread of education in the Lushai Hills and also the development of the Lushai language.

The two missionaries worked independently from 1894 but were called upon to move to Sadia to work among the Karbis and Abors in Assam. As a result they had to quit Mizo Hills leaving their work to Reverend D.E. Jones, a Welsh missionary who arrived at Aizawl on their departure on 31 August 1897. It also paved the way for the entry of a large number of missions of other denominations. But the duo came back again in an year and concentrated their energies on other philanthropic activities. They assisted the poor and oppressed hillmen and brought them before the authorities for redress. They helped the Lushais with medicine and books. Their hard labour and care made a lasting impression upon the minds of the Lushais. The Lushais held the *sahibs* in awe for a long time. But now they came in close touch with that dedicated couple and began to realize the 'virtues' and 'kindness' of the British people. The missionaries of the Welsh Mission who followed started their work from Aizawl and Lungleh. Members of the Welsh mission did their best to improve the lot of the Lushai people. They were more interested in charitable works than proselytizing. They were practical missionaries who realized that the time was not ripe for conversion. The Lushais were divided into so many clans and each clan had its separate language or dialect. But the dialect *Dulien* or 'Lushai' was the *lingua franca* among the majority of the clans. But the Lushais never attempted to codify the rules of language. The missionaries of Welsh Mission did this in 1897–8 by compiling a Lushai primer. This was so simple that a boy of ordinary intelligence could read fairly well after a fortnight's instruction, and in a month could write to communicate. But the Lushai primer compiled by the Welsh Mission used the method of transliteration adopted by Savidge and Lorraine in giving a written shape to the Lushai language. Savidge and Lorraine on the other hand were indebted to the writings of Captain T.H. Lewin. Lewin came in conflict with the Government of Bengal and gave up his lucrative job and dedicated his whole life to work among the Lushais. He wrote books on the Lushais to draw the attention of philanthropic people. It was he who first compiled a small dictionary of Lushai words. We have already mentioned that the Welsh Presbyterian Mission came to Lushai Hills under the leadership of Reverend D.E. Jones and Reverend E. Rowlands. The Welsh Mission established primary schools attached to the Church at Aizawl and Lungleh. After the missionaries had prepared the ground, the British administration established three schools for Lushai children around 1896–7 at Aizawl, Lungleh and Demagiri. But the standard of teaching was always higher in the Welsh Mission School at Aizawl. The students were taught Lushai through the English alphabet. The students of the age group 8–12 were higher in the rolls strength of the school than the age group 13–18. The London Baptist Mission Society set foot in south Mizoram in 1903. The Lakher Pioneer Mission originated in Mizoram itself in 1907 under Reginald A. Lorrain. Later the organization was renamed as Lakher Independent Evangelical Church. This private mission continued working will 1925. Very soon Savidge and Lorraine were joined by the latter's brother and his wife. The Lushai gradually found their presence reassuring as they provided invaluable medical and humanitarian services. But as far as conversion to Christianity was concerned there was little to be optimistic. The first conversion took place on 25 June 1899 under Welsh Presbyterian Mission when two men named Khuma and Khara were formally baptised. But after that, progress was again insignificant. The Census of 1901 recorded that there were only 45 Christians (0.05 per cent) out of 82,436 persons. By the Census of 1911, it went up only to 2,461 out of 91,204 (2.69 per cent). As can be seen up to 1911, the success rate of conversion was rather slow. The missionaries were quite frustrated. Then came the famine of 1910–11. It was over only by 1912. But the famine-relief work by the missionaries bore fruit. The next census recorded a 26 per cent increase in the number of Christians. In 1921, the head count of Christians was 27,720 out of 98,406 (28.16 per cent). In 1931, it went up to 59,123 (47.52 per cent) out of 1,24,202, which meant an increase of almost double the existing numbers. It is significant that in 1929-30 too there was another famine.

Adversities and Oppositions

loss of hope

The missionaries found that proselytization on the Mizos were 'hard and often disheartening' because 'the Lushais of that time had their vices and their virtues'.[29] They were said to be lazy, cruel, superstitious and very prone to drunkenness. It (Lushai religion) was their traditional faith and they were reluctant to relinquish it. Since the government had entered the hills so recently and since the heads of the government were whitemen, many Lushais took the missionaries as agents of the government.[30] The first missionary saw that it was essential for the spread of the Gospel to destroy this misconception. This took some time but at length the prejudice died out. The preaching was taken literally when the missionaries' spoke of being saved through the blood of Jesus some enquired what kind of magic would be found in this blood.[31] In fact, Reverend Jones' enthusiasm and the fact that on every possible occasion he spoke of Jesus puzzled many. He preached in the market, in the young men's community hut (*zawlbuk*) and often had services on Sundays in various places. The Lushais called him a 'mad white man'.[32] But when success came it was certainly not due to the missionaries themselves alone.[33]

Although the number of Christians were increasing in the hills, they were facing stern opposition from the chiefs.[34] There were many cases of persecution by the chiefs. There were hostile reactions from the parents and the public as a whole to their wards abandoning Mizo rituals and cultural symbols. In fact, their chiefs accused Christians of disobeying orders when they observed Sunday as no-work day. 'Most chiefs were hostile to the Christian movement, perhaps because they thought Christians were becoming a disruptive element in the normal village life. It was also possible that they saw Christianity a possible threat to their already limited authority as it involved acknowledgement of some one, Jesus, as Lord.'[35] As the number of Christians and seekers grew, hostility to Christians was also growing. The young preachers were often challenged by village strongmen (as was the practice amongst the Lushai to challenge visitors) to a bout of wrestling. Sometimes men used to try to compel them to drink *zu* (Lushai rice beer) and this happened to Reverend D.E. Jones himself. A bamboo cupful of *zu* was pushed in his face and almost down his throat. Soon, when converts began to appear, they too were persecuted. It was done in several ways. The Christians were sometimes beaten up during worship services and some of them were also injured.[36] Most of the chiefs had indulged in persecuting the Christians. The converts were forced to perform forced labour, fined unnecessarily, husbands were encouraged to batter their Christian wives and women were stripped and paraded naked.[37] Converts were expelled from village and made to leave their homes at midnight. Vanchhunga, one of the first evangelists in the north, had a list of five chiefs who persecuted Christians most. They were—Vanphunga of Zawngin; Thangkama of Sihfa; Lalzika of Buhban; Dorawta of Saitual; Lalruaia of Lailak. Chiefs also forbade giving food to Christians and denied them the right to cultivate. In some villages non-Christians refused to bury dead Christians, which was a real trial when there were two or three Christians. It was contrary to the traditional Mizo custom to refuse burial to anyone. Usually the Christian message was communicated through hymns and singing and it was spreading fast. So the opponents made a counter-attack using the same instrument—song and added dance to it, which the missionaries abhorred.

The slow growth in the number of Christians was also because the missionary's dislike of indigenous customs and rituals and attempt to introduce Western culture in its place which the tribals resisted.[38] At the very outset the Welsh Presbyterian missionaries (then known as Welsh Calvinistic Methodists) who were instrumental in pioneering Christian faith in the region made it clear for its members:

No undue haste however was made to baptize the converts. The standard of Church membership of the Mission field was set high and clear from the beginning. It was enjoined that every candidate should not only have renounced all heathen practices and lead a moral life, but that he must possess an intelligent knowledge of Christian principles, observe the Sabbath and abstain from all intoxicants.[39]

A senior missionary expressed the frustrations of the missionaries. D.E. Jones wrote in his autobiography,

We ourselves eagerly believed that the whole of that country would be evangelised despite all opposition. When we learnt of the success of Missions in Uganda, Korea, the South Sea Islands and other places. We hoped for something similar. But helpless our work seemed at times. . . . A missionary's faith and patience is tested when he sees a promising convert who begins to slow down, lose his enthusiasm and turns away from the Light. Our privilege is to pray for them. We don't know whether we shall find success despite many vicissitudes.[40]

The missionaries however had not lost hope by such hostility. They reminded themselves that 'Not by power, and not by might, but by my spirit, saith the Lord.'[41] Similarly the persecution of the early converts were seen as 'a cause for rejoiceing' because 'the blood of the martyrs is the seed of the Church'.[42]

The Famine: God's Strange Means

But all these lasted till 1911. Then came the famine that struck the Lushai Hills in 1911. The reports of its imminence had already been circulating among the people for the past years. But when it actually hit the area the people were devastated. Both the tribal and the missionaries turned their attention to the immediacy of the problem pushing aside other objectives for the time being. Reverend J.M. Llyod confirmed, 'It is significant to note that this ribald spirit (of opposition to missionaries) remained till the time of famine in 1911.'[43]

The missionaries had not witnessed the earlier bamboo flowering and consequent famine that devastated the Mizo Hills. They arrived later. But they had heard the stories. By 1908, *mao* bamboos had started to flower again and the tribals anticipated another famine in the next two to three years. Despite the slow rate of progress in conversion, the missionaries had by this time become very closely associated with tribal life. They learnt about the phenomenon of bamboo flowering and the ecological dynamics that caused famine from the Lushai. The missionaries shared the anxiety and concerns of the tribals of the impending danger and decided to come to their aid. The missionaries were already involved in humanitarian work in the Lushai Hills. They had set-up schools and provided opportunities for education. There were medical and health care facilities offered by them. Even though meagre, it had relieved many a common disease of the Mizos. The famine relief had been added to their list of humanitarian works, which they happily rendered. It was because 'the missionaries have always been faced with the truism that their success depended mainly upon their popularity with the people'.[44] Famine provided one such opportunity to gain popularity among hostile people. They started off with assisting the administration in relief. Significantly, during the famine the missionaries suspended their proselytization work even though that was their primary objective. The famine of 1911–12 and 1929 provided a great opportunity to the missionaries to introduce the 'human face' of Christianity to the Mizos. The famine was in fact described by the Missionaries as 'Gods strange means to spread his good news'.[45] The real famine year was 1911, though it started in 1910. The 'Flowering of Bamboo, Rats and Famine' took up a good portion of *Annual Baptist Missionary Society Report* for the year 1912. To quote:

But man was powerless against such a visitation (famine), and nothing that was done seemed to make the slightest impression on the great army of invaders (rats) which over-ran the land, and, in spite of everything, from the Chittagong Hill Tracts, right across the Lushai Hills, and away into the Chin Hills in Burma, the crops were destroyed wholesale.[46]

This was followed by the actual famine. It provided a golden opportunity to the missionaries to not only win over the chiefs but also extend their coverage. During the period they ceased all evangelical works including proselytization and concentrated on humanitarian work, which included arrangements to provide relief to the distressed. None of the missionaries had witnessed the famine of 1881. The 1911–12 *mautam* was the first famine experienced by the Lushai under British rule. The periodical flowering, seeding and dying of certain species of bamboo all over these hills was followed last autumn by an enormous number of jungle rats. But for the missionaries this was an opportunity of 'enhancing the saviour's kingdom'.[47] The missionary report stated that 'the giant spectre of famines have been spreading distress and sorrow all over this fair land but we have been spared the still more terrible experience of pestilence which at one time seemed to sweep the country and the trying times which we have been passing have strengthened our faith and have been the means of extending the saviour's kingdom'.[48]

The missionaries had plunged into famine mitigation in all their sincerity and resources. They were pained to see the 'proud' and hard working Lushai compelled by hunger to 'move door to door' and 'beg' for little food.[49] The Welsh Mission at Aizawl requested their headquarters in Wales to sanction financial assistance to meet the requirements of famine relief. Due to the famine the prices of all commodities went up. The wages of the servants and salaries of school teachers working in the mission schools had also risen. These were additional expenses to the normal relief expenditures, which the mission had to meet. The dearth of food made it impossible for the evangelists to travel as much as in former years. The same scarcity of food prevented them last spring from having the usual great 'gathering' of converts at which the annual baptismal service had taken place. The

Church membership decreased from 208 to 199 owing to death and migration.[50] But this was the occasion when the missionaries needed to step up their humanitarian work. The first task was to protect the crop from the invading rats. The missionaries tried all kinds of devices to keep off the terrible invasion. Fields were closely fenced in and traps of falling logs were set to catch any rat, which might attempt to enter. On one field alone 500 were caught in a single night by continually resetting the traps. In fact, it was quite usual to see people coming home with baskets of dead flattened rats on their back, which they had taken out of their log traps.[51] These would be dried and used as food, but it was hardly a diet to make up for lack of rice and most of them would be tired of eating rats, which at normal times would actually be a luxury. Fortunately a number of Lushais had some rice left over from the previous harvest. But they had the greatest difficulty in preserving it even when stored in their own houses. The people to be most pitied were those who from the beginning of the famine had no food at all in their houses. For such people, authorities imported rice from the plains of Demagiri. The missionaries themselves had been travelling throughout the autumn and winter about the country, helping the government and famine stricken people by superintending their distribution of food and collecting statistics on the condition of each household in the village with a view to further relief.[52] Scores of men and women who had no food to eat had been enabled to go down to Demagiri for a fresh supply of food by the loan of a pounds of rice apiece. Many others were provided employment in building, road making, construction of Churchs, forestry and gardening for food. The few who were unable to work were assisted with gifts of rice. The cheapest kind of imported rice, which could be had at Lungleh, was selling at Rs. 10 per maund (about 39 kg) and even at that price the supply available was limited. Owing to the famine, convent schoolboys had often had to leave their lessons to go and search for something to eat in the jungle. For several weeks the missionaries arranged a party every morning to hunt for supplies for the day collecting edible roots and leaves. In the month of May it became evident that rice would soon not be obtainable at Lungleh owing to a problem with transportation. An occasional supply kept appearing at Demagiri, a village at the foot of the hills, which was four days journey. It could not be brought nearer for want of men or bullock. So the church school-masters and boys were asked to go to Demagiri to carry the rice uphill. Simultaneously there had been great deal of sickness all around and in distant villages. Generally the missionaries would assist the administration in providing relief to the affected as they did not have enough resources to meet the entire relief expenses themselves. The administration was also dependent on the missionaries for the distribution of the relief material, as they did not have the manpower. Thus the two institutions complemented each other in famine mitigation. Alternately, the missionaries also helped the tribals in procuring relief in another way. As has already been mentioned, the administration had provided relief in the form of repayable loans. But to secure a loan, a tribal had to provide security either from the chief or a missionary. The chiefs were often reluctant to provide guarantee to too many commoners, in which case the missionaries had to step in. In other cases a missionary would sign a recommendation letter for a loan and send the applicant to the Superintendent.[53] The Superintendent had given permission to forward such deserving cases to which he would grant forty pounds of rice once a month.

Many old people and orphans come, as well as invalids and deformed. I send as many as I can to government for help but we have to help ourselves. They may have no place to stay or no cooking utensils. I have helped many to buy rice at cost price. Men who receive daily wages find that they cannot support their own families on what they receive. Some depend on jungle roots until they become ill and many don't want to beg. Many will never be able to pay back loans in five years.[54]

Writing about the famine ravages, the Annual Report of the Welsh Mission recorded,

The famine due to the growing crops being destroyed by rats, continued during 1912 in many parts of Lushai and caused much suffering and ill health among the people. Outbreaks of cholera in several districts accompanied it and carried off considerable number of victims. Many sad cases of destitute orphan children and other poor people had to be dealt with. Helped by the kind contributions towards the Lushai Famine Fund, opened by the Mission Directors and by other gifts we were enabled to relieve a good many sufferers.[55]

Some time the mission, starved of resources, was not in a position to tackle the widespread distress but continued their humanitarian work.

The Lushai has been suffering for some time from serious and very general famine. What to do is a great test of our faith. Many children, women, decrepit, blind and paralysed have come to the Mission Compound seeking help. Some have come from a great distance. From one village four decrepit persons were brought to Aijal by the chief as they coul not be supported by him. We have tried to help as many as possible of these poor people. Our resources at times seem to fail us. But so for our Heavenly father, the God of the widow and the orphan, has always remembered us and has listened to our cry. While I am writing this there is not a day's supply of money or food for the 50 people or so on this compound but we cast ourselves upon the Almighty father of our Lord and Saviour. He will provide, 'He that spared not His own Son shall freely with, Him give all things. If we can help them at their time of distress, there are great possibilities for the extension of Christ's Kingdom.'[56]

Thus the missionaries themselves felt,

In many ways we have been able to alleviate the want and distress around us and the gratitude of the poor people has been most pleasing to witness it has been a peculiar privilege to be living on the Lushai Hills this year and thus be able to help the poor in the hour of need. They have always looked upon us as friends and at such time as this poor especially find our presence a real source of real comfort and strength for they feel that they can come to us in their extremity.[57]

Like the colonial power, the missionaries too decided to use the opportunity to 'earn the goodwill of the people' which would facilitate their proselytization project. The method used was the humanitarian deeds. In other words, the missionaries too had their own politics during the crisis of famine. The objective was to succeed in their proselytization endeavour, through winning the gratitude of the people. The missionaries themselves admitted that the famine provided a good opportunity for them in the following way:

1. It (the famine) stopped the conquering power of the new heathen song which had drawn away most people from the Gospel during the past few years.*
2. It made the whole country free from rice beer for there was no rice.
3. It gave the missions an opportunity of witnessing through Christian Social Service.
4. It gave the Christians an opportunity to show their Christian virtue in such a situation.
5. The government was asking the missionaries to help in supervising the relief work, which entailed visiting most of the villages, which gave them the unique opportunity to fulfil not only the cultural mandate but the missionary mandate.
6. Their physical hunger made the people hungry for the Word of God and people were turning their ears again to hear the preaching of the Gospel.[58]

The missionaries were suitably rewarded for their work. Statistics showed that the Church was growing. By the end of 1912, churches and preaching stations were to be found in 80 villages, and 1,800 converts, of whom 552 were communicant members. The reward was all the more significant if it is taken into account that the missionaries had suspended their evangelical endeavours during the famine. However there were other events related to famine that took place in the Lushai Hills, which had contributed to the popularity of missionaries and the faith they were preaching. One of them was the controversy over the issue of slavery in Lushai Hills.

*The reference to the heathen song will be discussed in a subsequent section.

Slavery, Missionary and the State

The 1911 famine ravages were complicated by the controversy over the abolition of slavery in the hills of the Lushais. The issue was raised first in 1909 but gained prominence during the famine ravages complicating the situation. A non-conformist missionary named Dr. Peter Fraser, a physician by training took a keen interest in the institution of slavery as it existed in the Lushai Hills. He moved the government for its abolition the discovery since it was oppressive and violated the dignity of the people. The colonial administration was averse to the idea, as it feared any interference with the socio-cultural institutions of the tribals might incite an uprising. As it was, the British had quietened the fierce tribals only after half a century of bloody warfare. Instead they preferred to show the door to the missionary who was anyway becoming too popular among the tribal for his humanitarian concerns, which was another source of danger to the administration.

It was in 1884 that the first missionaries set foot in the Lushai Hills. The Welsh missionaries led by D.E. Jones followed them in as early as 1897. Dr. Peter Fraser along with his wife and another missionary, Watkins Roberts of Caenarfon joined them at the end of 1908. Fraser was a medical doctor by training and the missionaries of all denomination were very happy to have to him in the Lushai Hills, as there were no professional doctors there though a hospital had already been established.[59] Dr. Fraser himself subsequently built a well-equipped dispensary where there was accommodation for a number of patients who were provided suitable treatment. The doctor developed an immediate rapport with the tribals as well as with other missionaries. People came down to him from all parts of the Lushai Hills for treatment and in the first year of his stay he had attended about 10,000 people for various ailments.[60] The doctor had become so popular among the indigenous population, particularly the youth that the administration kept a close watch on him. The reason for the uneasiness of the administration was the radical nature of Fraser's interventions.

In course of his proselytization endeavour Fraser confronted opposition from the Lushai chiefs who would not allow his subjects to be converted to Christianity. These groups of people were called *bawi* and were under the control of the chiefs. The missionaries had already been facing hostility from the chiefs right from the beginning as the new faith threatened the power, and position of the chiefs over their people. Still the chiefs could do very little in preventing the people from embracing the new faith. But they could effectively stop the *bawi* as they were under their command and the *bawis* had no will independent of the chiefs. Enraged by the opposition of the chiefs and intrigued by the social position of the *bawis*, Fraser investigated the matter and concluded that the *bawis* were slaves and the institution was a form of slavery. Fraser resolved to fight the situation by asking for the abolition of the institution itself. This was in consonance with the policies of the mission to which belonged. Fraser tackled the situation in two ways. One, he liberated as many *bawis* as possible from their masters by paying a particular sum, Rs. 40 in this case, as required by custom as the price of liberation. Two, he moved his headquarters and the local administration for the total abolition of the institution arguing that slavery had already been banned by law in all British colonies including India.

In December 1909 itself, Fraser wrote to his headquarters in Wales about the institution and the urgency of its abolition not only because it was an abominable practice but also because it hindered the growth of Christianity in Lushai Hills. The headquarters permitted his intervention in March 1910 and issued a general circular to all the missionaries instructing them unequivocally not to condone slavery in any form. 'Do not condone slavery in any shape or form as we believe it to be contrary to the very essence of Christianity and to British Law and if it be proved that slavery is a part and parcel of the *bawi* system, that part of it should in our opinion be abolished at the earliest moment.'[61] In early 1910, an incident at the Chingchhip village aggravated the situation. The chief of the village called all the Christians and forced them to drink the alcoholic beverage, *zu*. As a result all the Christians of that village reverted back to their original religion. Fraser himself wrote of the incident, 'We had heard that all the Christians there had reverted to paganism. The chief there is very cruel. He recently compelled all the Christians to come together and his strong men forced *zu* down every Christian throat.'[62] The situation worsened during the famines. Since the chiefs were affected by the famine, their *bawis* were left unattended. Fraser took it upon himself to get the release of as many *bawis* as possible by paying their ransom money. But it was beyond his financial capacity to secure the liberty of all the *bawis* who contacted him for assistance. Hence he decided to move the administration to intervene in the matter immediately. The administration was not very happy with the inundation of the Mizo Hills by missionaries and the kind of intervention they made in the socio-cultural life of the tribal without taking any permission from the state. They did not like Fraser's interference in the social life of the Mizos. Such ventures would prove to be dangerous as it could anger the chiefs and incite a rebellion. Hence they pretended to be unconvinced that *bawi* was in any way a form of slavery.[63]

Representation of Slavery: Abolitionists *versus* Colonialists

To represent his case to the Government of India and argue in favour of its immediate abolition, Fraser prepared a report illustrating that the *bawi* institution was nothing but a form of slavery in the Lushai Hills. It stated,

> Lushai is country about the size of Wales. The people, about 90,000 in number mostly live in villages. Each of these villages is ruled by a chief who is generally of the Sailo family or clan. According to the old Lushai custom these Sailo chiefs have the right to have *bawi* or slaves.
>
> That this custom is really a system of slavery is evident from the following features: the slaves are bound to serve the chiefs for life unless ransomed by the payment of ransom money, generally forty rupees per family (£2 13s. 4d). Children of slaves are bound to serve for life, as also their descendents, generation after generation.
>
> When slaves move from one village to another they are still slaves for life to the new chief on the whole land they settle. By a new rule recently lay

down by Colonel Cole when Superintendent, the new chief is at once liable to the old chief for the ransom money. This is more clearly buying and selling of slaves than the old Lushai custom, under which the new chief was not liable to the old chief, but the slave changed his master without money being due from the new chief.

The *bawi* system is a system under which British subjects in Lushai are deprived of their right to liberty and justice is evident from a perusal of the following statements of slaves, evangelists, chiefs, missionaries and others.

Besides bondage for life other evil features are seen:

1. The inhumane separation of mother from her child
2. The separation of husband and wife
3. The separation of relatives
4. Intimidation, bodily hurt
5. Temptation to immorality and sin
6. Opposition to slaves becoming Christians
7. The selling and buying of people.[64]

Besides the feature of the institution of slavery, Fraser collected signed interviews from chiefs, slaves, former slaves, missionaries and responsible people as evidence of the prevalence of slavery in Lushai Hills and sent them over to all other missionaries. Once he was able to garner the support of few more to the cause, Fraser wrote to the administration to 'take action with the view of abolishing it . . . in the name of Lord Jesus Christ and in the name of our King Edward'.[65]

Major Cole, Superintendent of the Lushai Hills however was not convinced of the arguments by Fraser. Cole was already tensed over the radical activities of this medical missionary and his growing popularity among the indigenous people. Right from the beginning of colonial rule there was implicit tension between the colonial administration and the missionaries. The administration wanted the missionaries to confine themselves to the ecclesiastical activities but the latter often overstepped their limits and took part in reforming the tribal. The administration disliked the idea of the missionaries attempt to clothe the people whom they considered as naked tribal, opening schools and educating them, changing housing patterns and so on. All these amounted to detribalizing the Lushais whereas the avowed policy of the state was absolute non-interference in tribal affairs so that the tribal were conserved.[66] As far as Fraser was concerned, through the miraculous curing power of modern medicine Dr. Fraser had already become the darling of the tribals. They had stopped going to the charitable dispensary of the administration and instead came to missionary dispensary.[67] Fraser erected several new buildings during 1910–11 and liberated many *bawis* by paying the ransom money. He personally liberated at least forty *bawis* by paying their ransom money and even set-up two large hostels to shelter the liberated slaves.[68]

Major Cole was aghast at the idea of abolishing the *bawi* system. It not only necessitated serious interference with the historically evolved customs and traditions of the Lushais but also debilitating the institution on which the British Raj was established. Although the British were successful in subjugating the Lushai tribals, the administration was entirely based on the co-operation of the Chiefs. In fact, the British through the institution of the chiefs governed the Lushais. The abolition of the *bawi* system would mean the collapse of the privileges and power of the chief. The British Government could not afford another uprising from the Lushai chiefs whom they had subjugated after decades of warfare.[69] Major Cole called over Dr. Fraser and tried to convince him that *bawi* system was not comparable to slavery as understood in the West.[70] Fraser was totally convinced that *bawi* was slavery and threatened to move the imperial government and raise the issue in parliament through his good friend and then Chancellor of the Royal Exchequer at London, Lloyd George.[71] Major Cole, as representative of the government disagreed with Fraser's interpretation of the *bawi* system. He in turn cited the authority of Lieutenant-Colonel J.W. Shakespear, one of the earliest and most knowledgeable of the Lushai Hills superintendents who was well versed in Mizo language and their socio-political institutions. Shakespeare had written that 'among the Thados and Chins real slavery used to exist and men and women were sold like cattle. Among the Lushais this has never been the case. But there is a class known as *boi* who have been miscalled slaves by those ignorant of their real condition.'[72] The administration accepted Shakespear's views as the official position on the issue. The term *bawi* was also translated by Lorraine and Savidge the first missionaries in their first dictionary of the Lushai language as Slave or Retainer while others maintained that 'serf' would be the closer English rendering of the word *bawi*. The administration maintained that the *bawis* were not real slaves whereas most of the missionaries both in north and south Lushai Hills regarded it as a form of slavery. In this battle of self-righteousness the voice of the tribals themselves were lost. Both Shakespear and the missionaries were self-styled experts on the Lushais. There were no doubts raised about the correctness of either version by the respective groups. The administration had its reason not to go by Fraser's version, as it would mean the abolition of the institution, which could in turn threaten the stability of the Raj in these hills. Hence they found it convenient to uphold the view of Lieutenant-Colonel Shakespear.

The prevalence of slavery in a tribal society was not exceptional. In any manpower short economy, slavery has been the most elementary way of labour supply. After all, the Lushai economy survived on head-hunting and kidnapping. The primary objective of the head-hunting raids was to kill and obtain human heads for the mortuarial rites and kidnapping people to enslave them. These slaves could be from the neighbouring villages, tribes or the plains men living in the foothills. There were different kinds of slaves in Lushai society. According to the oral history of the Lushais, again recorded by the European administrators or missionaries, the slaves could be divided into two categories, the captive slave or non-captive slaves.[73] The captive slaves were those who were taken forcefully during inter-tribal wars, kidnapping raids or prisoners. The non-captive slaves were those who due to some distress bonded himself voluntarily to the chief. The captive slaves could be sold. The latter would be provided food and shelter by the chief against which the person had to perform labour services to the Chief. He could earn his freedom by paying a ransom in kind or cash. During the colonial time the British fixed it at Rs. 40, which was quite a huge sum for the tribals. These non-captive slaves could be further subdivided into three groups:

Chemsen slaves were those who committed serious crimes like murder and took refuge at the chiefs house to escape vengeance or punishment. The shelter came at a price: lifelong slavery. *Tulkut* slaves were deserters from the enemy tribes who defected to the victorious chief during war and consented to be a lifelong slave to him. *Impuising* slaves were commoners from within the tribe who due to poverty, distress or physical debility offered themselves as slaves to the chief in lieu of lifelong food and shelter. With the advent of the British, first two types of slavery ceased to exist. It was only the third type that remained, over which the controversy arose. The administration's opinion was that such a *bawi* was not a slave as he himself had volunteered to join the category. Secondly he enjoyed almost similar comfort and perquisites which other members of the chiefs household was entitled. Finally he could earn his freedom by paying the price of liberation. In reality however, a slave life was quite distressing and the so-called freedom he could achieve after paying the price was a utopia. Things had changed for better to some extent under the colonial administration but still the institution was as oppressive. The administration was fearful that the liberation of slaves would enrage the chiefs who could rise in revolt against the British rule. Hence it did not want to do anything to destabilize the newfound peace in the hills. Even the missionaries had agreed with him as it could jeopardise their proselytization project though they did not agree with the administrations position on the *bawi* system. However they decided in favour of maintenance of *status quo*. Dr. Fraser was thus left alone in the field.[74] Undaunted, he continued his crusade.

As the issue moved to the higher authorities, it became complicated. The Secretary to the Government of Eastern Bengal and Assam, B.C. Allen, acting on Major Cole's report favoured action against Fraser. He felt that Fraser's endeavour had produced immense discontent amongst the Lushais and if not checked could result in a general uprising.[75] He further observed that it was not for the missionaries to dictate which tribal customs should or should not be recognized, but it was the Superintendent who was responsible for the peace of the hills to decide whether time was ripe for effecting any change. Backed by the higher authorities, Major Cole threatened to expel Fraser from the Lushai Hills unless he gave a written assurance of good conduct. The undertaking required that Fraser would confine himself only to his medical and ecclesiastical works and not interfere with tribal customs and traditions.[76] Fraser refused to do so even when his church headquarters at Liverpool requested him.[77] Fraser instead took the matter to the Anti-Slavery Society at London to rouse British public opinion about it. They wrote to Edward Montague, Under Secretary of State for India, to take immediate action to procure the freedom of the slaves in the Lushai Hills. Following the uproar, John Gardine raised the issue in the House of Commons on 12 June 1913. Montague in his reply to the question raised by Gardine went by the local administration's version that no slavery existed in Lushai Hills and due to the threat of political de-stabilization caused by Fraser's intervention in the affairs of the tribal, the government had banished the missionary from the Lushai Hills.[78]

Fraser left Lushai Hills but won a moral victory as the government decided that henceforth (1) the use of the term *bawi* would be discontinued as far as possible, (2) claims for *bawi-man* (price of freedom) should be treated exactly like the claims which any Lushai, not a chief might advance against any persons to whom he had given board and lodging, (3) claims against any one family of *bawi*s should not exceed Rs. 40, (4) that the chief could not force any *bawi* who had left him to come back without recourse to judicial procedure and *bawis* were free as well to leave his chief if he was discontented in the chief's household by an appeal to the law court or requesting the administration to pay his freedom amount, and (5) any other question in connection with the *bawi* system would be decided according to universal Lushai custom as binding on all Lushais, without any distinction between the chiefs and the ordinary people.

But all these happened after Fraser had left the Mizo Hills. On his refusal to sign the undertaking given to him, the administration asked him the leave the hills immediately. Both Fraser and Major Cole were called to Shillong but neither changed their position on the *bawi* issue. The final decision was that neither man should stay on in Mizo Hills. Fraser's residence permit was withdrawn and Cole was transferred to Manipur state. In September, Fraser suffered a serious and prolonged bout of enteric fever. He had nearly died but he recovered well enough to sail to Britain leaving the Mizo Hills on 26 November 1912. The departure of Peter Fraser coincided with the cessation of famine. But both had left their imprint: Famine, its devastations; Fraser his deeds. Fraser had in course of his stay of four years treated about 25,000 Mizos. This was out of a population of about 92,000 (1911 census). Fraser himself wrote about the famine months, 'The medical work has increased in amount during the nine months, January to September, the average number of outpatients case treated was about 2,000 a month. The number of inpatients was at times greater than could be accommodated in the dispensary, so that some had to be admitted into the orphans' and schoolboys houses for a while. Several came for treatment from a long distance, over 100 miles in some cases.'[79] Fraser left behind patients, almost one-third of the Mizo population who remembered him and the Christian Mission gratefully. What is most significant is the concluding part of the Annual Report of the mission which stated, 'It is a cause of deep thankfulness that most of the patients, after a stay in the hospital, have willingly given their names as believers in Christ and also tried to lead others to Him.'[80] His colleague remembered, 'Dr. Fraser developed the medical work considerably. People came to him from all parts of the land . . . Mizos soon saw that the Christians lived longer and there were fewer deaths among them than among non-Christians . . . we were amazed at the confidence people had in missionary medicine. . . .'

Fraser was of a generous nature and during one year he supported scores of youth. They were given education and afterward they were to go far and wide throughout the hills to spread the Gospel. He placed eight of them as teacher-evangelists in impoverished villages. They were appointed as full time teacher in schools, by the Mission after he left.'[81] Despite being embroiled in the *bawi* controversy during the famine, Fraser did not neglect his patients. In fact, he worked more during this time of crisis. He had also liberated at least 40 slaves and arranged for their livelihood before he left. Not just these 40, he had left behind a great mass of grateful *bawi* population for whom he fought a lonely battle. The reports of his struggle against the administration spread across the Mizo Hills through word of mouth after he left the hills. Later missionaries reaped the harvest that Fraser had sown. But the work though which Fraser directly furthered the cause of propagation of Christianity was the creation of a band of youths called the *kraws sipai*.

Rallying the Youth: The Soldiers of Cross

Peter Fraser was not only a radical preacher but very creative in his ideas. It was precisely because of this that he was immensely popular among the tribal youth. Fraser was quick to realize the potential of this raw youth power, which he decided to utilize in the cause of Christianity. In course of his stay in the hills Fraser could easily see that the stumbling block to the objective of Christian missions were the older generations, which stubbornly stuck to their old religion, customs and rituals. It was the autocratic chiefs who resented Christianity as they stood to lose their prestige and power from its spread. Hence they created all kinds of obstacles to discourage the missionaries and prevent proselytization. Under such circumstances it was the youth who were the hopes of missionaries. Fraser therefore turned his attention to the Mizo youth who like other young men were curious and eager about new ideas and institutions. As he developed a rapport with them, Fraser conceived of the idea of a band of youth who would be dedicated to the cause of Christianity and would supplement the work of the missionary preachers voluntarily. He called them Soldiers of Cross or *kraws sipai*.

As referred to earlier Fraser had liberated many slaves from their masters by paying their ransom money. He had also constructed two large hostels to provide shelter to them. While staying in the hostel they would often gladly turn out to help

Fraser when he went on his preaching tours, especially to carry his harmonium. They also joined him in singing hymns, forming a chorus. With this group of young men Fraser formed *kraws sipai* in 1911.[82] Khuma, the first Mizo Christian was a prominent member of this group. Inspired and supervised by Fraser this band would go from village to village spreading the message of Christianity. This group of zealous Christians travelled together over the hills, carrying with them their own supply of food and preaching the Gospel wherever they went. They were unpaid voluntary preachers.[83] Over the north-eastern border, in the hills of Manipur state, in the village of Senavawn, where the kinsmen of the Mizos lived they had heard about the developments taking place in the Mizo Hills and sent a message to the Welsh Mission if it could send somebody there to explain the new faith. Since there were no missionary to spare and Manipur being a sovereign Hindu state, permission of the government was necessary. Jones decided to send the Soldiers of Cross over. Reverend D.E. Jones' strategy was twofold: since they were not official missionaries they did not need permission; being Mizos they could explain the words of Gospel better. Thus by the end of 1911 five youths aged between 16 and 26—Vanzika, Savawma, Taisena, Luaia, and Thanghurha—visited the Senavawn and surrounding villages of Manipur.[84] During the famine this group also assisted the missionaries as well as the administration in relief operation and at the same time propagating the objectives of the mission. The Annual Mission Report recorded,

> This year (1911) they made a *jhum* or cultivation and were successful to some extent in growing rice and other crops. They have formed a band, which they call *kraws sipai* (Soldiers of Cross). Their aim is the salvation of their fellow-men and they go out to the villages far and near to tell of their achievements. They endeavour to maintain themselves and to help other members of the band. One of them, Hranga, has been accepted by the Presbytery as an evangelist, to labour in the South-Eastern portion of North Lushai. In response to pressing requests from several villages, six young men who volunteered for the work have been helped to open and maintain schools. The aim is to seek and win souls for Christ and very encouraging results have been reported from time to time.[85]

The Politics of Revival Festival

The greatest event of the 1911–12 famine *vis-à-vis* Christianity was the organization of the Great Revival. 'Revival' has been given to mean, 'a time of extraordinary religious awakening or working up of excitement especially accompanied with extravagance'.[86] It was further defined as 'the working of inner spiritual experience with a sudden illumination—being born again in Christ—coming after the consciousness and repentance of sin'.[87] The evangelical enthusiasm was the outcome of a religious revival movement, which originated with John Wesley and Whitefield, the two prominent Methodists. The revival stemmed from overall socio-economic changes and spiritual bankruptcy, which swept the decadent English society in the wake of Industrial Revolution.[88] Consequent upon the revolution the growing polarization between the English industrial capitalists and the working classes resulted in acute tension in society. The increasing exploitation and deprivation of the working classes created rising cultural, psychological and moral depression among the lower echelons.[89] It was among these depressed labouring men that revivalist events found its greatest adherents. Revivalist events brought spiritual distraction and thus neutralized the tension by calming down the revolutionary temper of the people. They generated 'a new enthusiasm and intense moral earnestness coupled with a deep concern for the unsaved'.[90] The history of Christianity is inextricable from Revivals. This is especially true of the Welsh. A number of Welsh missionaries were working in north-east India and they used their home experiences in this part of the world as well.[91] This was the background of the Revival in the Mizo Hills. It was important because,

> The phenomenal success of the Gospel preaching in Mizoram was brought about through revival movements. The whole people have been converted to Christianity within the space of 50 years is nothing short of exceptional. The Gospel preaching began in 1894 and revivals affecting mass conversions started after 12 years from which time repeated waves of revival swept the land so that the end of Second World War in 1945 saw the whole land embracing Christianity to the extent of total abandonment of their old religion.[92]

Revival has been an inseparable part of the history of Christianity in Europe from the early twentieth century. The great revivals of 1904, 1905 and 1906 were well known. The revival events in Wales were not just an extreme spiritual experience but also an expression of its distinct national identity.

Henry VIII followed by Queen Elizabeth had accepted the Reformation movement of the sixteenth century wholeheartedly in England. As a part of England, the Welsh also accepted it but it did not make any deep impression on them. Neither the English Bishops spoke the Welsh language nor did they visit the Churches of their diocese. They functioned by collecting information through intermediaries. The history of war and hostility with the Anglo–Saxons haunted the Welsh and they craved for their own national identity. The attitude of the Anglo–Saxon priests further alienated them. They needed a religion in which they could not only have the fervour and national expression but also a satisfyin, spiritual experience. It was Howell Harris, a Welsh nationalist who on being refused priesthood in the Anglican Church founded the Welsh Church in 1737 through a frenzied spiritual movement called Revival. It was not just the establishment of a separate Church but a national movement. Seceding from the English Church was symbolic of the secession of Wales from England. It took a religious form.

> At the centre of the movement was preaching, which moved along well defined lines. The message 'The Cross of Christ' presented in two parts; first a powerful denunciation of sins with threats of Divine judgement. Second the free gift of God's forgiveness, salvation and eternal life through Jesus Christ. As excitement helped preaching, preaching excited people out of the round of their daily life and habits causing them to change their lives.[93]

The movement went on during the rest of the eighteenth century continually. In the process there was reformation of the religious and social life of the people, publication of the annotated Welsh Bible in 1770, educational reforms by men like Thomas Carlyle and the final separation from the established Church resulting in the formation of the Welsh Calvinistic

Methodist Church at the close of the century. Thus the Presbyterian Church of Wales started under the above name when its missionaries arrived in north-east India. Each revival was preceded by a dull period or crisis in public life, which would then be rejuvenated by the organization of revival. What is relevant here is the striking coincidence that these crises had with a famine or epidemic that frequently hit Wales.[94] Famines and plagues were frequent in Wales and Ireland. Whenever these adversities devastated the people their helplessness against calamities made them realize the power of a supra-entity and drove them towards God[95] which was the secret of the popularity of revival during which 'several people might be praying at the same time while others were singing and others telling things while most of the people were crying'.[96] The most known revival event in Wales was in 1904 under the leadership of Evans Roberts. It was called the Greatest Welsh Revival 'because of its extent as well as its intensiveness. With great force and influence it spread throughout the country and it was believed that among the people living in Wales at the time only very few would be left outside its influence'.[97]

The Welsh Revival had succeeded in sweeping the Wales of United Kingdom with religious fervour. Observing such success, the Welsh Mission in the Mizo Hills, which was going through a series of frustrations in its proselytization efforts, tried to replicate the revival in this part of the world too. 'As elsewhere the growth in the Church in Mizoram was slow until the revival (*Harhna*) came. I had longed to experience the revival as it had taken place in Wales during 1904–5.'[98] Accordingly the head of the Welsh Calvinistic Methodist Church in Mizo Hills sent its Mizo converts to a revival meeting in nearby Shillong but it did not meet with much success. On their return they however displayed the frenzied emotional and spiritual outburst, which the missionaries termed as the first revival.

In Wales during the revival we knew of people who had fallen into trances. This happened in Mizoram. To some it may seem accidental. I can only report what I actually saw some fell into such a coma that we did not know if they were still alive. At first we were frightened by this for it looked they had crossed into the world of the spirits. But we were assured by those who had witnessed similar incidents in the Khasi Hills that they were in no danger. When they awoke, they spoke of visions they had seen and of future events which came to be realized.[99]

The Welsh Mission was aware that the revival was most successful in Wales during crises like famine and epidemic. So when the 1911 famine struck the Mizo Hills they decided to organize the second revival. This was considered necessary since, in the previous years the Church had been facing an extreme low with opposition from the chiefs, and a new cultural revivalism of the animist faith among the Mizos. It was felt that the famine-time adversity was best to organize another revival by which the tides could be turned in favour of the Church. As soon as the famine showed signs of decline, a revival was organized, this time in the Mizo Hills. 'The 1913 revival began in Champai within a few miles of the Chin Hills in Western Burma. . . . Later it came to Durtlang, six miles to the north of Aizawl and to Aizawl a week later.'[100] The result was stupendous. Not only the slow growth and opposition to the Church was countered but there was tremendous increase in the number of converts. During the second revival the Christian population increased to 16,935. The 1911 census recorded 3.12 per cent of the population as Christian but the next 1921 census it went up to 30.02 per cent.

There was another famine in 1929. The last revival organized in the aftermath of the last famine had seen huge success. There was en masse conversion in certain areas. It prompted the Welsh mission to organize another revival event after the fresh famine.

Not only famine; there were accompanied calamities as well in 1929. Under heavy rains portions of hillsides slipped down here and there in varying size and dimensions confirming the tribal belief that impending bamboo flowering had always brought other disasters in its wake. 'Landslides of 1929 were exceptional in that they were widespread all over the country. We are told that whichever way you turned your eyes, you could see hill slopes dotted with landslides. The cause of the calamity was said to be the unusually heavy rainfall that year. Incessant downpours for days specially during 1 to 10 June (1929) backed by violent exploding rumbles of thunders which shook the earth caused the landslides. Crops were destroyed resulting in the inevitable famine. The year was marked in history of Mizo Hills as the year of Great Landslides (*Minpui Kum*).[101]

The calamities prompted spiritual resurgence. As the missionaries organized relief they also took advantage of the public gathering that took place during relief operations. They used it for publicizing the powerlessness of men against nature and the strength of prayers in such time. The vulnerable people joined the frenzied revival meetings. Stirrings of revival began to appear towards the end of the year in and around Champai. Several Church leaders from Aizawl attended a Presbytery meeting at Champai in early 1930. Lingering features of revival such as ecstatic dancing and singing were features of these meeting. As the excitement grew it spread to most villages of the hills. 'During those years, revival and for that matter Christianity was gaining in popularity and public acceptance in the land, more strongly in eastern, southern and south-eastern parts of north Mizo Hills and the scattered villages of south Mizo Hills. Society with its very culture was seriously shaken when people embraced the new religion en masse.'[102] In the annual report of the Presbyterian Church of Wales for the year 1919–20, when a revival festival was organized, it was recorded,

> The first place must be given to the great revival which so deeply affected the life of the churches in Lushai. It not only brought great blessing to those who were already believers, but it was also the means of adding more than four thousand to the membership of the Church. In villages where previously no Christians were found, there are now churches meeting regularly for worship and in villages where there were but few Christians; there are now large numbers, eager also to preach the Gospel and to help to gather in their neighbours. Bands of young men and women are travelling from village to village, preaching and exhorting and there is a spirit of gladness in the churches which makes it very enjoyable to visit them.[103]

The result was there for all to see. In 1931, the next year, the percentage of Christian population went up to 59,123 (47.52 per cent) out of 1,24,202, which meant an increase of almost double the existing numbers. However the exact increase was visible only in the next census. In 1941 it went up to 98,108 (64.00 per cent) out of 1,52,786. The Christians were already in majority by this census. In fact, the greatest increase in a single year happened in the years 1933 and 1934. In 1933, there was an increase of 6,229 to the total tally of Christians, and in 1934, it was 7,578. This was a massive increase in a population of less than 1.5 lakh.

The organization of revival festival was Welsh experience which the missionaries experimented with in Mizo Hills. The other such experiment which was put into practice in the region was the rice collection system. This was more directly connected with famines. The Welsh missionaries felt that if the tribals were induced into the habit of saving it would help them in times of distress like famine. Since they had little cash available with them it was found wise to induce them to save in kind, in this case, rice. But this collection was not just for the 'rainy day' but also to cover the expenses of church organizations. Hence it had to be stored with the church. The credit for the experimentation with this idea goes to the wife of D.E. Jones. The Welsh missionaries in Shillong had introduced it in the Khasi Hills which Mrs Jones had heard about. The practice was to store a fistful of rice everyday for the church by every member of a family. At the end of the month, this collection, which would roughly become about 2 kg would be presented to the church to be sold. It brought a substantial sum which was then used to cover the growing expenses of the church in running its affiliated schools, dispensary and so on. The money was used to construct new chapels, school structures and church buildings. The resource was also used to mitigate famine related shortages.

Cultural Revivalism and Indigenization of the Church

The famine and the revival festivals organized in its aftermath facilitated the spread of Christianity in two other ways. Firstly, it stopped the onslaught of the cultural revivalism that had swept and almost doomed the cause of Christianity in the Lushai Hills on the eve of the famine. Secondly, by the sheer force of its cultural vitality it compelled the foreign church to adapt to the cultural institutions of the region. Once the church condescended to do it, it did not remain a foreign religion anymore. It was just another version of the indigenous religion, which encouraged people to join. This was the secret of the success of the church in the Lushai Hills in the twentieth century. Both took place against the backdrop of the periodic famine.

The colonization of the Mizos by the white men had been a turbulent event in the history of the tribe. The advent of the missionaries compounded the situation. The missionary attempt at conversion to Christianity actually involved a multi-pronged attack on tribal society and culture. Firstly it branded tribal life as 'savage'. Secondly it subjected every aspect of Mizo life—from their nakedness to rituals—to severe denigration. Even their songs, dances, sexual norms, gender relationship were singled out for criticism. Thirdly, they were encouraged to give up their traditions, which they held dearly for time immemorial, and accept a new faith. An alien human form was presented as their god in preference to their own *pathian*, who was neither visible nor imaginable. The colonial state had already snatched their political freedom and dislocated their economic life. It made them lose their power and image of a mighty people capable of terrorizing the plains people. The missionaries only supplemented it by attacking their cultural life. There was a sense of defencelessness and alienation from such attacks. A British administrator himself depicted the situation as:

The advent of the British form of government and control for a time certainly paralysed people. This was inevitable. Raiding excursions by Lushai had been countered abruptly by the permanent incursions of a power and a might far beyond the full comprehension of innate Lushai. All through the history of Lushai relations runs strong evidence of the respect and desire felt by people for power. The British occupation of Lushai Hills marked the presence of a power hitherto unforeseen and unimagined. The world of the Lushai was staggered, bewildered. Within their land, the Lushai people soon found two powerful contacts were at work. One was the government in the personality of the political officer, later the superintendent, Lushai Hills, and the other were the missionaries, who had come on the heels of the British conquest. The former aimed at securing peace, law, and order while the latter aimed at converting the Lushais from their animist beliefs to those of the Christian religion as interpreted from their standpoint. . . . Against these varying contacts the Lushai had no equipment on which to fall back for strength except their traditions and the stories of their grandfathers.[104]

Initially they ignored the missionary propaganda followed by fierce opposition. When even that was failing against the aggressive proselytization efforts of the missionaries, a cultural revivalist movement was started, which gained momentum around 1908. Certain traditional songs and dances were revived and aggressively performed throughout the Mizo Hills by the tribals. The Mizo culture, which was losing ground suddenly found itself revived. It immediately caught the fancy of the people. They participated in it with renewed vigour. It was an attempt by the tribals to fight back and come on top of the new culture that was being introduced by the combined power of the British and the missionaries. Knowing that the missionaries did not like the Mizo way of song and dance as it involved drinking alcohol, free mingling of men and women and sexual innuendoes, it was performed more aggressively in front of the missionaries and the recently converted tribal Christians. This was an attempt to spite the white men and isolate the converts from the mainstream of tribal community. It was a cultural revivalist movement of the Mizo popularly known as *puma zai* movement. It was believed that *puma zai* was originally a prayer song addressed to god in sacrificial worship by the Biate tribe who were early occupants of the northern parts of the Mizo Hills. *Puma zai* therefore was a song in praise of the god Puma.[105] Another theory has it that Puma was a village bard who had composed catchy two-liner songs about the traditional life and culture of the Mizos. It had melody, humour and colloquial language. It was also participatory. People who were performing it could use their own ideas too and create a new song in the same tune. Thus common people not only participated in it but also used these songs to comment on contemporary events, which included the advent of the British, the works of the missionaries and their own situation. The song or chant was complete in two simple lines—the first ending with Puma while the final three syllables of the second line was repeated.[106] The rise and fall of the cadence was in a regular rhythm but the beat of the syllables in the line varied for more of less in length. The very simplicity of the tune and naiveté of the composition made the song easy to repeat. The following are few examples:[107]

Our village water-spring is a good spring Puma,
For we spied belladonna taking bath, taking bath

Our village blacksmith is an evil blacksmith Puma,
For he can't do a welding properly, iron welding

Our village priest is an evil priest Puma
For he cant wait sacrificial meat well cooked, well cooked.

Thus the song could run on and on and cover any and every aspect of contemporary life. Subsequently *puma zai* underwent transformation and modification in the hands of Mizo poets especially Hmarawna also known fondly as Awithangpa. It basic form also changed. Instead of two liners it became three liners and the name Puma was dropped to replace *thanglam zai*. The new name came from the change in the form of its performance. Formerly one person dancing in the centre circled by singers staged this dance of drinking bouts. But with *puma zai* all the singers became dancers to transform the performance into a public dance. Thus it began to be called *thanglam zai*—song of public dance.[108]

As the news of *puma zai* spread even the inaccessible areas sent their representatives to go and learn the performance and bring to its village. The chiefs were most enthusiastic supporters of this movement. They took active part in sponsoring and spreading the movement throughout the length and breadth of Mizo Hills. It was a prolonged festival all over the hills of feasting, dancing and drinking. Although *puma zai* was not religious in its nature, soon it incorporated some ceremonial celebration with *ran lu kima ai*—full rounds of animal heads. The song and dance movement spread like 'flaring quickness of cotton wool flakes and the intensity and extent of enthusiasm was such that the like of which had never been known.'[109] The *puma zai* festival was successful firstly because it revived interest in the beleaguered Mizo tradition and culture. Through *puma zai* the Mizos rediscovered tradition and were reassured that their culture was unassailable. Secondly, it countered the invasion of an alien faith and culture represented by Christianity. It made the propagation machine paralysed. The missionaries would generally sing hymns to start a preaching meeting. But with *puma zai* people refused to attend those meetings. The *puma zai* also used to taunt and ridicule the missionary preachers. They were made an object of common contempt. Thirdly, the *puma zai* movement succeeded in bringing the new converts back to the folds of community and its tradition. J.M. Lloyd a Welsh missionary described the situation thus:

> One of the severest tests came in 1908, when there was a sudden resurgence of heathenism. An old Lushai tune was set to new words and became immediately popular. The words were generally in praise of a great village chief (chiefs then were persecutors of Christians). It was reputed and by many, to have been sung by a jungle spirit. It spread like wild fire to all parts of the hills. Amazing manifestations of feeling accompanied the singing—almost as though the revival were parodied. Great feasts were held during which the young men and girls danced in ecstasy. These demonstrations were made in every village. The travelling preachers complained that preaching was a burden. The gospel was losing ground and no one wanted to listen to it. The cause of Christ seemed doomed in Lushai.[110]

But the famine of 1911–12 came as a providential help to the missionaries. The food shortage, starvation and threats to life weakened the movement. As the crisis intensified, people withdrew from festivities in search of more pressing needs of food. Since the missionaries were organizing famine relief they were no more seen as aliens. In fact, people rushed to them for help. The contact with the missionaries was also necessitated by the fact that they were the link between the administration and the people. Both in the north and south Mizo Hills 'the salvation (for Christians) from the influence of *puma zai* came ultimately in the form of *mautam* famine in 1911 and 1912 which effectively put a stop to all public feastings and entertainments'.[111] The missionaries gladly accepted the opportunity. 'It is significant to note that this ribald spirit and the popularity of this song [*puma zai*] remained until the time of famine in 1911. It was only then that preaching in the villages began to be a little easier. The rice famine made the people hungry for the word of God.'[112]

As the famine was over and both the people and the missionaries got back to their daily life, a gradual change was noticed in the attitude of the missionaries toward tribal culture. Although the missions were happy with the growth of Christianity in the hills they did not forget the near defeat in face of a cultural revivalist movement. In retrospection they felt that Christianity as was practised in the West would not perhaps acceptable to the tribals. Exuberant dance and songs were inseparable part of their culture and any attempt to curb them would only create further hostility. In fact, if they were absorbed as part of the Christian culture in this part of the world, it would pave the way for mass acceptability. Acculturation and indigenization was after all not new to Christianity. It has been a part of its history. The indigenization of the church in Mizo Hills was therefore conducted in two ways. First, more and more tribals were employed as evangelists and spreading the network of church across the hills. By network it meant churches in every village, clinics and charitable dispensaries, schools and educational institutions in every interior of the region. The second was accepting tribal feasts, dance and songs as part of the Christian festivals. The revival festival for example could not be held without the tribal song and dance. The English hymns were too slow for such ecstatic activity. In every church therefore a floor space was allotted in the central area for dancing, which was between the communion table and the central area. Dancers moved out there and danced in circles to the end of the singing. Once a person started dancing he or she had to carry on till the end of the song which was repeated from the beginning for many times. Since the performance was in the churches, it was religious in nature and under the supervision of the priests. It never took the form of sexual suggestions which was the case in tribal dance ceremonies. A church historian confirmed the situation as, '. . . the missionaries and early Christians were almost all opposed to the observance of traditional festivals . . . because the songs and dances were perceived to be either sexually suggestive or celebrations of war and head-hunting . . . because of their association with old religion . . . excessive dancing and drinking involved'.[113] He assessed the change in the following words,

> In due course the attitude of both missionaries and local Christians changed towards many of the things that had previously been frowned upon. . . . One example is provided by developments associated with the first great revivals in Mizoram. While the Welsh missionaries strongly opposed the use of drums and dance among the Christians, their use was introduced in the revivals and have remained an important part of Mizo Christianity ever since.[114]

This was very significant as it entailed a strong institutionalized religion like Christianity succumbing to local cultural pressures and allowing itself to be adapted to tribal cultural features. It, in fact, was not tribal conversion to Christianity but to use the words of Susan Bayly,[115] Christian conversion to Tribalism. The result was indigenization of Christianity which rendered it acceptable. It was confirmed by a Mizo church historian as well,

> Underlying this numerical increase was a growing sense of ownership and spontaneity, a process of indigenization. Previously Christianity was looked upon as something foreign, imported and inculcated into the society by the White people and their native helpers. And it would have taken quite a lot more time for the chiefs and their councils to accept the new religion had it not been for their public who embrace it in mass conversion beyond their control. But now Christians through their experience of the work of the Holy Spirit gained new confidence that Christianity belonged to them too and not the 'Mission' only and that the Church itself was their own. This sense of ownership as well as of belonging to one another in the Church was a very new thing and became a binding force that gave them strength to grow up in spontaneity and maturity.[116]

Renewal of Famine

The next famine was due in 1929. The experience gained during the last famine (1911) made it easy for the government and the missionaries to provide for the next. As early as 1924, the bamboos started flowering indicating the imminence of the famine as predicted by the tribals. The missionaries began to act immediately to tackle the impending crisis. They knew, church alone could not counter a calamity of this magnitude. In fact, it was the state which had to bear the major share of the responsibility. What the church could do was to assist them with ideas and assistance. On 17 January 1924, Reverend J.H. Lorraine wrote to the district administration:

I am taking the liberty of writing to you regarding the expected *Thingtam* famine and although I have no connection with the government I insist that expressing my own opinion as to the means which might be employed to successfully counteract the effects of such a visitation will not be unwelcome to one who like yourself has the welfare of the Lushai people so much at least.[117]

Lorrain then talked about the sufferings of the people and his experience during the last two famines and suggested certain actions from the end of government.[118]

1. That the government should pass order to make thrift compulsory so that a certain portion of harvest was put aside by each of the Lushai household from now on. The proportion could be 1/6 or 1/5 so that by the time famine arrived, each household would have enough savings of rice.
2. The farmers should also be encouraged to cultivate little more land than they did to produce that extra bit. Each house could have a small rat-proof granary of its own to save the rice.
3. The government should introduce the use of Liverpool virus which was said to exterminate such vermin by causing a deadly epidemic among them and reduce the number of rats. Also, people should be then prohibited from eating the rats as they would be virus affected.

The government however felt that the ideas of missionary were excellent but unworkable because 'to begin with it would be a serious interference with the liberty of the subjects and that the measures proposed although for their ultimate good would be greatly resented'. Generally the Mizo's annual harvest did not last more than six months of his food requirement. The bumper crop of 1922 had enabled only about 60 per cent of the community to have a full year's rice. The administration felt that to enforce saving of grain as advocated by the missionaries, would cause great hardship on the remaining 40 per cent who were already short of food. In other words to save for future, the tribals had to starve in the present. Again to enforce Lorrain's proposal of punishing offenders on these grounds would be very unfair.[119] As far as the use of Liverpool virus was concerned the government sought the advice of Colonel Hodgson, Director of the Pasteur Institute who opined that (1) no virus was effective in decreasing rat population and (2) rat population increases enormously with even a temporary increase of available food supply for rats, as a female rat was capable of producing a litter within six weeks of its birth.[120]

The government instead recommended:

1. The publication of article in Mizo *chanchins* (newspaper) and also make Circle Officers interpret and try to impress on all chiefs the necessity of putting aside little grains for a rainy day.
2. Destruction and preventing the increased supply of food to the rats, i.e. the bamboo seeds by burning the bamboo jungle.
3. An intensive poisoning and rat trapping campaign throughout the country. The Barium Carbonate poison and the wonder traps of French pattern could be provided by the government.[121]

But by 1929, the famine hit the hills and it was time for the state as well as the philanthropists to start working. The experience of the previous famine came in handy. The effort of the government as well as the church leaders in convincing people about the need to save bore fruit. The good harvest of the past few years enabled them to save enough. Coupled with that the wholesale slaughter of jungle rats over a considerable period by the people at the initiative of the government and church leaders reduced the menace considerably. 'This greatly eased the situation and although there have been considerable hardship and sufferings, actual starvation has been avoided except by rats themselves. These have died in such numbers that as on previous occasion the survivors are not numerous enough to prevent a good harvest from being reaped in a few weeks time'.[122]

On the mountains, in the hairpin bends of Kolodyne River there were villages, which nearly always escaped the savages of the rats. Two Christian chiefs who were brothers ruled the two largest of these. At the beginning of the famine one of them had such an abundance of grain that he was able to make arrangements with government to supply such grain at a fixed price to villages scattered over a wide area. This was a great block of hundreds. The other brother was also able to help a large number of families who came for rice. The pastor residing there wrote to Lorrain,

> I want to tell you what a difference the love of Christ has made to this part of the country. During the great *thingtam* famine of nearly 50 years ago this village was spared the ravages of the rats and had abundant grain while most of the country was starving. Much the same thing happened again this famine but the behaviour of the people has been strikingly different because of the Gospel. They being afraid of feeding people who flocked to the village becoming too numerous blocked the paths and even killed those who persisted in getting through. Now because of the light of Christ's love they want to help all who come to them and the destitute and maimed are being fed without any payment. This shows how desirable the Gospel is and how it can change the very nature of man. I feel sure the supporters of the mission will be glad to hear this.[123]

The 1930 famine created a lot hardship for the people but the counter offered by the church was so much that in the same year the Baptist Mission could hold a grand revival. A revival was generally held after a depression in the proselytization movement. It was organized by respective churches in the form of festivities. A revival ensured resurgence in the movement. It resulted in new enthusiasm, and new recruitment. The positive response to the missionaries after the famine from the Mizos prompted the church to quickly organize a revival so as to reap the harvest of this positive attitude. The missionaries already had a similar experience of appropriating a natural disaster to enhance their proselytization projects, elsewhere in the north-east. It was during the earthquake of 1897, which devastated the Khasi Hills that the missionaries used calamity to indulge in humanitarian activism, earn the goodwill of the people and thereby attract them to the Christian faith.[124] A similar situation was reported from the Indian state of Andhra Pradesh as well which being a coastal region had been prone to devastating oceanic cyclones. The cyclones not only brought catastrophe to the people of the region, the failure of monsoon would often create famine situations. The failure of the south-western monsoon caused a serious famine in the state in 1876–8. Though the famine was widespread in the whole Madras Presidency, its impact was most severe in the Andhra region. It affected the economic and social life of the Andhra population because it dislocated agriculture and industry. Prices showed a steep upward trend and crimes were on the rise. The government entered the scene in a big way to provide relief to the affected people but it was the philanthropic work of the missionaries, which was noteworthy. During this period of distress the missionaries practically stopped all evangelical works to participate in relief activities. They successfully secured huge foreign grants from the English and American mission establishments and distributed them in the interiors of Andhra region. Various relief measures were carried out. Following the decline of the famine ravages it was noticed that the Christian population had almost doubled. A study of their relief operations suggested that famine was one of the links in the chain of events such as mass conversions, establishment of schools and hospitals, and the actual famine relief in kind provided to the affected people. The doubling of the Christian population in such a short period was a result of these endeavours.[125] A similar situation was observed in the Cachar district of Assam, which was frequently inundated by the river Barak during the monsoon creating havoc in the lives of the people. Although Christian missionary activity started here early, there was no discernible rate of conversion among the dominant Hindu and Muslim communities that inhabited the valley. However, during flood ravages the situation was quite a contrast. As usual the stationed missionaries would plunge into relief activities and there would be a growing interest in the Christian religion. Such floods were there in the years 1915, 1916, 1929, 1936, 1946. After the 1929 flood the mission reported,

> The last year was one of unusual activity as the result of the flood which so tragically devastated the area. During the months, which followed the ordinary routine work had to be abandoned. This was no unmixed evil . . . the deep appreciation of the people of the little we were doing for them were but touching the fringe of their necessities; the gladness with which they welcomed us everywhere and the hospitality often form unforgettable incidents. Never had we come into closer communion, and never had they so shared in our lives. This intimate contact with what Christianity suggested in time of crises has certainly left a deep impression. More than once we were invited into *madrasa's* and as we stood there barelegged and barefooted, and comically besplashed with mud, there was no mistaking the earnestness with which the Muhammedan boys and their teachers listened to our spiritual story.[126]

In the past the famine years in Mizo Hills there was a decline in the number of school children and church-goers. There used to be a large number of backsliders as well. But 1930 proved otherwise as can be seen in the following figures (Table 4.1).[127]

The report specifically mentioned the impact of famine relief by the missionaries. The famine thus had significant impact from the missionary point of view.

In the colonial period itself when an evangelist researched on why the Mizo became Christians he found the following set of answers from the tribal:

1. They did not need to sacrifice to the demons anymore.
2. Christianity meant salvation from all ills and suffering as *pathian* (Jesus) had power over all evil spirits as had been proved.
3. Freedom from costly and numerous rituals and sacrifices to be offered to the evil spirits.
4. Christianity was the way to get healing from sickness without sacrifice to the demons or rituals.
5. Free entrance to *pialral* (heaven).[128]

TABLE 4.1: GROWTH OF CHRISTIANS IN MIZORAM

	1929	1930
istian c community	10,398	11,209
Church members in good standing	3,728	4,059
New converts	451	603
Backsliders	411	248
Backsliders reclaimed	300	345
Sunday schools	90	95
Sunday school scholars	5,997	6,199
Sunday school teachers	261	312
Public learning to read in Sunday schools	939	1,172
Sunday school scholars able to read	2,546	2,763
Believers baptised	340	384
Catechumens being prepared for baptism	481	574
Deaths	230	219
Births	417	479
Bamboo churches	64	73
Days labour given free in building and repairing churches	3400	4,280
Financial contributions	2,957	2,311

The church's influence made a great difference to the people's attitude towards the famine. It was usual for the Christians to give a tenth of their crops as free will offering to the church. Due to the famine the previous year there were practically no crops to reap. In spite of this, a sum of Rs. 258–8 *annas*–10 *paise* had been handed in and as there was good surplus over from the year before the evangelists were not only supported but there had been enough to enable them to have an additional famine allowance. This free will offering thus collected was also used for famine relief. The new social attitudes were often most clearly seen and most impressive in times of crisis. The Mizo revival of 1913 for instance and a rapid growth of the church, which followed it has been attributed to the social attitude of Christians during the preceding bamboo-famine:

The notable difference between this and previous Lushai famines lay in the willingness of the Christians to share with their less prosperous brethren—both Christian and non-Christian. . . . During previous famines the prosperous villages (a number of villages usually escaped the plague of rats) used to erect stockades to keep out the famine stricken. Some of the hungry ones in desperation would try to storm these defences to obtain food. They were usually killed in the attempt. But during the famine of 1911-12 great brotherliness and generosity was shown. The non-Christians admitted that this was due to the influence of the Gospel and it made a great impression on them'.[129]

After the famine of 1911–12, there was a rise in the number of Christian (Tables 4.2–4.4). In fact, proceeding to 1911–12, there was a steady decline and setbacks to the proselytization process. After the famine, preaching in villages become easier. The people were already grateful to the white missionaries. The tiding over of the hardship of the famine was celebrated in the Church premises with the white men. The missionaries found the time ripe to strike. The rat famine made the people hungry for the word of god and the flower of revival was kindled again in 1913. This happened to be a special year in Lushai history for several reasons. The effects of the great famine of 1911–12 were still felt by the people. The brotherly spirit shown by the Christians during the famine had made a profound impression on the non-Christians. It was in this year that the Aizawl Christians had completed building the Mission Veng Chapel. It was a large building with the best material they could find had been put into it and most of the labour had been given voluntarily by the members. 1913 was also the year when the first Lushai pastor was ordained. All these helped the growth of converts in these hills.[130]

During the famine, children were the worst affected. 'Pagan' as well as Christian parents often collapsed in their efforts to obtain food for their children. Dozens of these orphans were brought to the missionaries who cared for them in spite of their own straightened circumstances. Government officers organized the import of rice from the plains and issued loans to the people. Collections were made in Wales to send to those who were starving in the Lushai Hills and this proved a great help. The widespread suffering sobered all the inhabitants and robbed the pagan singing and dancing of all its zest. The spiritual temperature began to

TABLE 4.2: CHRISTIAN STATISTICS IN NORTH MIZORAM (PRESBYTERIAN CHURCH 1899–1983)

Year	Christian Community	Communicant Members
1899	12	7
1900	15	5
1901	24	8
1902	40	34
1903	36	28
1904	57	19
1905	90	36
1906	167	74
1907		
1908	420	145
1909	626	334
1910	949	464
1911	1,723	552
1912	2,455	851
1913	4,776	1,665
1914	5,839	1,931
1915	7,168	2,805
1916	8,579	3,087
1917		
1918		
1919	17,838	6,418
1920	21,171	
1921	22,108	8,006
1922		
1923		
1924		
1925	34,893	15,220
1926		
1927	35,577	15,660
1928		
1929	35,880	16,817
1930		
1931		
1932		
1933	53,153	20,803
1934	59,566	20,850
1935	63,872	22,994
1936	64,983	28,882
1937	66,399	29,510
1938	66,945	29,864
1939	70,175	31,188
1940	72,173	32,077
1941	74,987	33,327
1942	76,657	34,069
1943	78,535	34,904
1944	80,584	35,815
1945	83,858	37,270
1946	87,617	38,941
1947	90,658	40,292
1948	93,103	42,290
1949	98,602	43,823
1950	1,00,513	44,672
1951	1,02,280	45,457
1952	1,03,405	45,958
1953	1,05,034	46,682
1954	1,0663	47,406
1955	1,08,141	48,063
1956	1,07,798	49,220
1957	1,15,625	51,016
1958	1,15,764	51,450
1959	1,20,707	56,375
1960	1,25,293	
1961	1,31,589	57,700
1962	1,31,504	58,108
1963	1,37,112	60,769
1964	1,41,064	63,482
1965	1,44,516	64,248
1966	1,48,132	64,571
1967	1,48,188	64,898
1968	1,52,848	68,412

TABLE 4.2 (contd.)

Year	Christian Community	Communicant Members	Year	Christian Community	Communicant Members
1969	48,942	26,068	1977	1,82,543	88,807
1970	50,411	27,154	1978	1,90,008	93,644
1971	50,411	27,593	1979	1,97,231	98,062
1972	46,468	25,227	1980	2,07,032	1,03,746
1973	46,991	24,753	1981	2,09,133	1,06,350
1974	52,235	30,097	1982	2,14,431	1,09,455
1975	52,235	29,820	1983	2,18,503	1,12,828
1976	52,426	29,073			

Source: C.L. Hminga, *The Life and Witnesses of Church in Mizoram*, Aizawl, 1987.

change and people listened with avidity to the preaching of the word. The travelling preachers began to be welcomed once more in villages. But the notable difference between this and the previous Lushai famine lay in the willingness of the Christians to share with their less prosperous brothers—both Christians and non-Christians.[131] In their report of the Lushai Hills for 1899–1900 Reverend D.E. Jones and Reverend E. Rowlands found that 'although the natives are always suspicious of other *sahibs*, the *Mizo sap* (missionaries) soon gained his confidence. Some of them say that they would like all the sepoys and officers to leave but they would wish the missionaries and the shops to remain.'[132] The cumulative result of the humanitarian works conducted by the missionaries to provide relief to the victims of recurring famines have been summarized by the report of the Mission itself.

> Whatever feeling of resentment may have lingered in the hearts of some of these people against those who have occupied their country in order to prevent a repetition of their head-hunting raids upon the peaceful inhabitants of the plains. The famine must surely dispel it. For there are hundreds who would have starved to death this year but for the *kindly* help rendered by government in bringing up many thousand of maunds of rice to supply their needs.[133]

In fact, in 1927, at the farewell meeting of D.E. Jones, a white missionary who was leaving Mizo Hills, both the Christian and

TABLE 4.3: CHRISTIAN STATISTICS IN SOUTH MIZORAM (BAPTIST CHURCH 1902–1983)

Year	Christian Community	Communicant Members	Year	Christian Community	Communicant Members
1902	45	4	1936	18,224	7,039
1903	125	13	1937	19,463	7,515
1904	259	36	1938	19,343	7,196
1905	314	68	1939	20,036	7,417
1906	304	86	1940	21,084	7,560
1907	368	108	1941	21,678	7,840
1908	500	136	1942		
1909	608	152	1943	20,108	8571
1910	805	182	1944	23,108	8571
1911	1,130	208	1945	25,232	9548
1912	1,544	199	1946	27,357	10,525
1913	2,647	348	1947		
1914	2,739	447	1948		
1915	2,772	714	1949	31,079	12,133
1916	2,686	854	1950		
1917	3,108	918	1951		
1918	3630	969	1952		
1919	3,670	1,017	1953		
1920	4,790	1,218	1954	37,760	16,226
1921	5,583	1,559	1955	37,876	16,423
1922	7,820	2,059	1956	38,183	18,886
1923	8,105	2,773	1957		18,886
1924	8,770	3,198	1958		
1925	8,965	3,335	1959		20,367
1926	9,720	3,354	1960	44,303	
1927	9,935	3,491	1961	43,744	21,362
1928	10,031	3,478	1962	40,614	19,085
1929	10,398	3,728	1963	47,678	23,056
1930	11,209	4,059	1964	47,753	22,957
1931	12,125	4,452	1965	48,786	20,713
1932	13,380	4,846	1966	49,447	24,014
1933	14,815	5,792	1967	47,296	23,679
1934	15,980	6,019	1968	47,209	25,500
1935	17,449	6,448	1969	48,942	26,068

TABLE 4.3 (contd.)

Year	Christian Community	Communicant Members	Year	Christian Community	Communicant Members
1970	50,411	27,154	1977	52,929	29,354
1971	50,411	27,593	1978	66,447	29,032
1972	46,468	25,227	1979	52,829	30,229
1973	46,991	24,753	1980	57,469	33,339
1974	52,235	30,097	1981	57,726	33,071
1975	52,235	29,820	1982	60,007	35,694
1976	52,426	29,073	1983	60,007	35,694

Source: C.L. Hminga, *The Life and Witnesses of Church in Mizoram*, Aizawl, 1987.

TABLE 4.4: CHRISTIAN STATISTICS IN SOUTH MIZORAM (MARA CHURCH 1899–1973)

Year	Christian Community	Communicant Members	Year	Christian Community	Communicant Members
1899	12	7	1942	76,657	34,069
1910	1	1	1951	7,560	
1911	2	2	1961	9,244	
1921	97		1971	12,937	
1931	910		1973	13,238	6,800
1941	4,866				

Source: C.L. Hminga, *The Life and Witnesses of Church in Mizoram*, Aizawl, 1987.

non-Christian chief were vying with one another to show their respect and gratitude to the departing missionary. A thankful Khamliana, one of the seniormost chiefs addressing a gathering of about 3,000 people said,

> We are all by this time glad that the British Government has come to our Country. Our inter-tribal wars and quarrels are over. The boundaries of the chiefs territories have been satisfactorily fixed . . . we have been safeguarded from famine . . . we are equally grateful to the Welsh Mission for coming here. . . .[134]

In the famine report of 1929–30, the evangelists ended with a significant statement, which the subsequent history of the Mizos would substantiate, 'The magnitude of change that has occurred in the Lushai is not easy to explain. . . . An astonishing, of profound and radical nature has occurred. . . . A new nation has been born.'[135]

Politics of Humanitarianism

The opening up of India for evangelical projects encouraged a number of trusts and missions, which sponsored young missionaries to the frontier areas of India where 'savage' hill tribes were not only threatening the stability of the Raj but undesireably also living outside the boundaries of the 'civilized' world. One such mission was the Arthington Mission, set-up by an English millionaire to tame the head-hunting tribals through conversion to Christianity in the distant Mizo Hills. It is curious that 50 years back an English paternalist, David Scott, who did not believe in changing the tribal life or institutions, sought the help of missionaries to convert and civilize the Garo tribes in the same region to prevent them from head-hunting practises in British territory. Arthington was prompted to do the same from far away England when he heard of the Mizo tribes murdering a fellow English planter and kidnapping his six-year old daughter to the jungles.

The missionaries therefore landed on the hills with a preconceived notion that the people who they were to 'civilize' were 'ugly savages'. The project stood for an ultimate transformation of tribal society. They directed their political interest in securing the legal protection of Christian converts, the supervision of inhuman rites such as *sati*, infanticide and head-hunting, discontinuation of British patronage to temples, mosques or religious festivals. The most important feature of their evangelical belief was that the human character could be suddenly and totally transformed by a direct assault on their mind.[136] They demonstrated that the Hindu divinities or tribal deities for that matter were absolute monsters of lust, injustice, wickedness and cruelty. In short the religious system of either was one great abomination.[137] So were the tribal gods and divinities. The evangelists were out to rid them of these primitive customs and religions.

The hostility of the tribes to any attempt at conversion, opposition of the chiefs to be replaced by an alien 'White God' (Jesus) and trespassing of the 'half-baked' human beings (whitemen) into the cultural domain of the tribal prevented any headway in the evangelization project of the missionaries. But nature intervened and a disaster happened. The famine provided the missionaries with a 'god send' opportunity. Evangelism was temporally substituted by humanitarianism. The distressed people were provided food, work, substance and aid. This rendered the religion and image of a vital practical religion in the minds of the tribal as against their own, just the way the missionaries visualized. The politics of humanitarianism paid off as those who earlier shied off the missionaries for fear of being deculturized, came back to them. Interestingly, while the tribals were deculturized during this encounter, the missionaries themselves had immersed in tribal culture. They used their mutual proximity to learn the Mizo language, share their food habits and participate in their festivals. It helped them shed their alien and hostile image. Once the closeness was established there was no great opposition to proselytization efforts. Every successive famine was followed by the organization of a revival festival, which reinvigorated the dwindling number of Christians. The accomplishment of task was not easy as the evangelists had to make certain adjustments. For example they believed that salvation was attainable only through refuge to the prophet for which knowledge of God was necessary. The attainment of that knowledge was possible only through an understanding of His revealed words—the Bible. Both the Methodists and the evangelist therefore concentrated on securing minimum education as a prerequisite for conversion, at least sufficient for a person to read and understand Bible.[138] But the Arthington Mission, which sponsored them, refused to finance the education of the tribal. The missionaries therefore had to arrange alternative source to establish schools. The Census of 1901 indicates that only 26 tribal professed the Christian faith; even in

1911 the number of Christians were very few in the Lushai Hills. The Lushais changed their faith *en bloc* in the 1930s of the twentieth century. Although the proselytization was not very satisfactory at the beginning of the British rule, the influence of the clergy in the Lushai life was overwhelming. The missionary teachers, doctors and nurses were all committed evangelists. A large number of European ladies of the Welsh Mission also joined the missionary enterprise.

The chiefs began to realize the value of education and sent their children to these schools. The wealthy chiefs were persuaded to send their wards to better institutions at Shillong, Silchar, Chittagong or Sylhet. Not only the children of the chiefs, but the children of the Lushai commoners also began to attend the schools in the Lushai Hills. It is to be remembered in this connexion that the contributions of the missionaries far surpassed the contributions of the government. They not only supplied the best teachers but also spent a lot of money on that account. Before 1897–8 the government spent nothing on that head. In 1897–8 the administration spent an insignificant amount of Rs. 333 on education. In 1898–9 the amount was increased to Rs. 902 and in 1899–1900 the expenditure incurred on education was Rs. 1,793. There is no doubt that the missions spent more money on tribal education than the colonial administration. A later colonial administrator regretted this decision[139] of the early administrators as he was able to see the difference it made; the tribals were transformed according to the ideas and image of the missionaries which was a contrast to that of the administration. As the bulk of the funds for the education of the Lushai people had been generated by the missions, it was only natural that the school teachers had been selected with due regard to their standing within the Lushai Church. It tended to make the teachers preach the Gospel than teach the basic skills. In course of their work they had developed a band of young Mizo followers. These lads were between the ages of 14 to 16. In fact, the evangelizing was done by these Lushai youth, and this was the essential clue to the fact that hardly anywhere in whole of the vast continent of Asia has the Gospel been spread more rapidly and effectively. In the Lushai Hills, it was always the youth who came to Christ first while the old struck to their old faith and opposed Christianity.

Secondly, to make the 'heathens' believe in the new god they had to be differentiated from the tribal gods and evil spirit. They did it by demonstrating the expensiveness of the elaborated tribal rituals and their viciousness. They also took recourse to modern medicine which could relieve a tribal of his simple ailments immediately whereas the tribal priests and his rites failed. It is another matter the tribal gods like *pathian* or *lalpa* was transposed with Jesus, which made in easy for the tribal to accept Christianity.[140]

Thirdly, the tribal customs and festivals such as uninhibited mingling of the two sexes, breaking out in songs and dances at every opportunity, consumption of alcoholic beverages and regular community feasts which the missionaries of civilization ardently disliked, were now appropriated and accepted as part of tribal Christianity. A great sacrifice indeed.

To come to the political economy of evangelization, the project itself was a political decision. It held the politically self-righteous position that the tribals in particular and Indians in general were in need of the civilizing hand of the West. To the evangelist, the hand of God was nowhere more surely seen then the miraculous subjection of India by a handful of English. Power carried with it an awful responsibility and duty—the salvation of India's heathen millions.[141] Yet the missionaries did not fail to see the connection between civilization project and commerce.[142] Commercial and missionary opinion agreed upon the fundamentals of the Indian problem and its solution. Together they generated colonial policy of nineteenth-century liberalism. This was the policy of assimilation and anglicization. The alliance between the missionary and free trade enthusiasts did not occur in 1813 but from the 1820s they were rapidly merging.[143]

The significant political undertone of the missionary movement was the alliance of the utilitarian and the evangelist in following the policy of assimilation and endeavour to transform the colonies in its own image. Even the tribal in a distant frontier land were not spared and sought to be changed to Englishness not just in dresses and religion but also in moral values. Utilitarian principles had thus reached the far away mountains from the shores of England.

Lastly in the evangelist project, the missionary's humanitarian concern had a particular objective. Like its colonial counterpart it believed in earning the goodwill of the people so as to made them to susceptible cultural *conquest*—a term which the missionaries actually had used in north-east India. Success did not elude them. The politics of humanitarianism paid dividends to such an extent that even a decade later in post-colonial India, when a similar famine struck them again, the grateful Mizos recalled the benevolent humanism of these white missionaries when the post-colonial political dispensation turned a blind eye to their distress.

The colonial administration completely paralysed the tribal's inner coping mechanism in fighting a natural calamity, which were inextricable parts of their life. By arranging relief, it made them completely dependent on the administrative machinery. The dependence was so extreme that the tribals were unable to think that they could manage to cope with calamities on their own. Gradually such dependence covered other areas of life like food, shelter, and subsistence. As the colonial administration was entrenched, most Mizos wanted employment in either administration or church institutions so that he could easily earn his livelihood and did not to have to go back to agriculture. 'Local educational practice soon gave rise to the belief that education and Christianity were the passport to 'salaried jobs', relief from the wearisome toil of cultivating a hard land. Education had constituted a means to a dead end—the salaried post. It has beckoned young Lushai towards distant lands and ideas rather than towards the land of their birth, the land of their future. The salaried post has been termed a dead end, because it has so often marked the cessation of further real effort.'[144] It completely escaped their thoughts that the British administrative machinery on which they were so dependent was an import of last two decades only. They had been fighting their battle against calamities on their own before the white men came along. Thus the dependence on the colonial state was so complete that capitulation was total; neither did they anymore think about fighting for freedom nor did the British have to worry about it any longer. This was the reason that when the British decided to withdraw from India, the tribals were in a state of shock. They did not know what they should do if the British were leaving? What was their future? There were so many questions and no answers.[145] While decolonization had created a sense of liberation elsewhere, in the hills of north-east India the mood was pensive. Such was the dependence on giant colonial machinery. The Mizos did not even remember that just 50 years back they had fought a fierce protracted war against the same colonialists and was able to resist their advance for almost four decades. What the missionaries did by their humanitarian philanthropic work was to consolidate this dependence. For the missionaries it was far easier to cement the dependence because unlike the government officers they stayed for years in the same place and same position in course of which he developed a personal relationship with the people.[146] The missionaries thus had an overwhelming power over the lives and ideas of the tribals. A British administrator in the Lushai Hills sarcastically called them the 'white chiefs' replacing the original chiefs in power and prestige.[147] In fact, unlike the officers, a missionary became a part of the happiness and distress of the tribal and thereby developed an interdependence based on emotions, which is why the tribal were traumatized by the eventual departures of missionary personnel. It is interesting that though the missionaries worked independently and were often in conflict with the colonial administration on issues of change, their work only complimented the imperialist project. The British government was the new Order and Christian missionaries through their work helped people to adjust and accept to the new Order.[148] They changed the tribal perception that the colonial administration was an alien government and made them feel as if it was their own. While the British made the colonial conquest of the Lushai, the missionaries, by transforming them morally and culturally, consolidated the conquest.

NOTES

1. Andrew Porter, *Religion versus Empire: British Protestant Missionaries and Overseas Expansion 1700–1914,* Manchester and New York: Manchester University Press, 2004; Gerald Studdert-Kennedy, *Providence and the Raj: Imperial Mission and Missionary Imperialism*, Sage, 1998; Stephen Neil, *Colonialism and Christian Missions*, London, 1968; Laldena, *Christian Missions and Colonialism: A Study of Missionary Movement in North East India with Particular Reference of Manipur and Lushai Hills 1894–1947*, Shillong: Vandrame Institute, 1988, p. 4.
2. Laldena, ibid., p. 7.
3. Arthur Mayhew, *Christianity and the Government of India, 1600–1920*, London, 1929, p. 50.
4. Eric Stokes, *The English Utilitarian and India*, Delhi: Oxford University Press, 1989, p. 14.
5. Ibid.
6. Francis G. Hutchinson, *The Illusion of Permanence: British Imperialism in India*, Princeton: Princeton University Press, 1967, p. 3.
7. Ibid.
8. Stokes, op. cit., p. 34.
9. E. Daniel Potts, *British Baptist Missionaries in India 1837: The History of Serampore and its Missions*, Cambridge: Cambridge University Press, 1967, pp. 174–95.
10. Stokes, op. cit., p. 35.
11. Reginald A. Lorrain, *5 Years in Unknown Jungles for God and Empire*, Liverpool: Lakher Pioneer Mission, 1912; rpt. Gauhati: Spectrum, 1988.
12. The issue has been debated in Robert Eric Frykenberg, ed., *Christians and Missionaries in India: Cross Cultural Communication Since 1500*, Studies in the History of Christian Missions Series, London: Routledge/Curzon, 2003, pp. 1–32.
13. Refer to Bishop Horne and Charles Grant's view cited in Stokes as note 4 above.
14. See for details Sajal Nag and Satish Kumar, 'Noble Savage to Gentlemen: Discourses of Civilization and Missionary Modernity in North East India', *Contemporary India*, Journal of Nehru Memorial Museum & Library, Delhi, vol. 1, no. 4, October–December 2002, pp. 113–28.
15. Suhas Chakravarty, *The Raj Syndrome: A Study in Imperial Perceptions,* Delhi: Penguin, 1991, *passim.*
16. Ibid.

17. Ibid.
18. See the description in Lorrain, op. cit. A'so, Reverend D.E. Jones, *A Missionary's Autobiography 1897–1927*, tr. from Welsh by J.M. Llyod, Aizawl: H. Liansailova, 1988.
19. H. K. Barpujari, *The American Baptist Missionaries in North East India, 1836–1900: A Documentary Study*, Gauhati: Spectrum, 1986, p. 13.
20. Ibid., p. 20.
21. J.M. Morris, *The Story of our Foreign Mission*, Aizawl: Synod, 1930, p. 77.
22. Jones, op. cit., p. 1.
23. Rev. P.H. Moore, 'Need of a Native Ministry', in Papers and Discussions of the Jubilee Conference of the American Baptist Missionary Union Held in Nowgong, December 18–29, 1886, Guwahati: Spectrum, 1997, p. 15.
24. J.H. Morris, *History of Welsh Calvinist Methodist Foreign Mission*, Aizawl: Synod, 1930, p. 54.
25. J.M. Lloyd, *On Every High Hill*, Aizawl: Synod, 1984, p. 24.
26. H. Yule, 'Notes on the Khasi Hills and People', *Journal of Asiatic Society of Bengal*, vol. 12, pt. 2, July–December 1894, p. 612.
27. K. Thanzauva (comp.), *Report of the Foreign Missions of the Presbyterian Church of Wales on Mizoram 1894–1957*, Aizawl: Synod, 1997, p. 11.
28. Rev. V.L. Zaithang, *From Headhunting to Soul Hunting*, Aizawl: Synod, 1981, pp. 11–16.
29. J.M. Lloyd, op. cit., p. 24.
30. Ibid.
31. Ibid.
32. Ibid., p. 25.
33. Ibid., pp. 27–8.
34. Rev. C.L. Hminga, *The Life and Witness of the Churches in Mizoram*, Aizawl: Baptist Church, 1987, p. 61.
35. Ibid.
36. Lalsawma, 'Four Decades of Revivals: The Mizo Way', *A Gospel Centenary Souvenir*, Aizawl, 1994, pp. 39–42.
37. Ibid.
38. Nag and Kumar, op. cit.
39. Morris, op. cit., p. 91.
40. Jones, op. cit., pp. 33–5.
41. *Report of the Lushai Hills, 1899–1900* by Jones, op. cit., p. 9.
42. *Report of the Lushai Hills 1909–10* by Rev. Dr. P. Fraser, p. 42.
43. Lloyd, op. cit., p. 24.

44. A.G. McCall, *The Lushai Chrysalis*, London, 1949; rpt. Kolkata: Firma KLM, 1977, p. 203.
45. Rev. Zairema, *Gods Miracle in Mizoram*, Aizawl: Synod, 1978, p. 8.
46. *The Annual Baptist Missionary Society Report, 1901–38*, Aizawl: Baptist Mission Society, 1993, p. 81.
47. Ibid.
48. Ibid.
49. Letter by Mrs Jones, Aizawl, 1912, 8C, 27, 318 NLW, CMA.
50. *Annual Report of the Baptist Mission Society*, South Lushai Hills, Assam, 1912, p. 81.
51. Ibid., p. 86.
52. Ibid.
53. Orform to Williams, Aizawl, 1 January 1912, 8-A, NLA, CMA, 27, 318.
54. Ibid.
55. *Annual Report of the Lushai Hills 1911–12*, in Thanzauva, op. cit.
56. Ibid., pp. 47-8.
57. Ibid., p. 90.
58. Ibid.
59. Jones, op. cit., p. 49.
60. Ibid.
61. *Foreign Department, External Affairs Proceedings*, 11 September 1910, nos. 5–21, p. 5.
62. Peter Fraser, July 1910, cited in Lloyd, op. cit., p. 154.
63. H.W. G. Cole to Fraser, 23 February 1909.
64. 'The Bawi System of Lushai, Assam, British India', *Letter sent to Colonel Cole, Superintendent of Lushai Hills through Mr. Von Morde, Assistant Superintendent*, December 1909, NLA, CMA 27, 318, File V, The Bawi system in Lushai: Dr. Fraser's Case and Letters 1911–13, Aberystwyth, Wales, UK.
65. Fraser to the Superintendent of Lushai Hills. January 1910.
66. Nag and Kumar, op. cit.
67. Jones, op. cit., pp. 50–1.
68. Llyod, op. cit., p. 152.
69. Acting on the Superintendent's report B.C. Allen, Secretary to the Government of Eastern Bengal and Assam argued that Fraser's propaganda had produce the greatest discontent and that his course of action, it not checked might result in a general uprising in the hills. B.C. Allen's Letter to Secretary to the Government of India, Foreign Department, 4 February 1911, p. 7.

70. Major Cole to Fraser, 23 February 1909, NLA, CMA 27,318, File V, p. 25.
71. Fraser to Major Cole, 22 February 1909, NLA, CMA 27,318, File V, p. 24.
72. J. Shakespear, *The Lushai–Kuki Clan,* London, 1921; rpt. Kolkata: Firma KLM, 2002, p. 45.
73. Ibid. Also see Laldena, op. cit., pp. 71–2.
74. Initially all missionaries of all denominations supported the cause of Peter Fraser. But as the administration threatened to withdraw its permission to the missionary to stay in the Mizo Hills, they had to withdraw their support. Even the Welsh Mission headquarters requested Fraser to ignore the issue. But Fraser decided to continue the crusade.
75. B.C. Allen's Letter to Secretary to the Government of India, Foreign Department, 4 February 1911, p. 7.
76. Draft Submitted by Major Cole and Expected by him to be Signed by Dr. P. Fraser as a Condition that he Should Return as a Missionary to Lushai, 16 December 1910, NLA, CMA, 27,318, File V, pp. 28–9.
77. W.E. White, 'The Relations of Missionaries to the Government and their Own Mission Councils, in *Indian Witness,* 4 September 1913, NLA, CMA, 27,318, File V, 1911–13.
78. *Parliamentary Questions on Slavery (Lushai), Answer to Sir John Gardine's Question,* no. 101, 12 June 1913.
79. *Annual Report of the Lushai Hills 1912–13,* in K. Thanzauva (comp), *Reports of the Foreign Mission of the Presbyterian Church of Wales on Mizoram 1894–1957,* Aizawl: Synod, 1997, pp. 51–2.
80. Ibid.
81. Jones, op. cit., p. 50.
82. Llyod, op. cit., p. 152.
83. Ibid.
84. Ibid.
85. *Annual Report of the Lushai Hills 1911–12,* in K. Thanzauva, op. cit.
86. Willian Geddie, ed., *Chambers Twentieth Century Dictionary,* Edinburgh: W. & R. Chambers, 1964 rpt., p. 944.
87. Laldena, op. cit., pp. 12–13.
88. K.P. Sengupta, *The Christian Missionaries in Bengal 1793–1833,* Calcutta: K.P. Bagchi, 1971, p. 7.
89. Eric J. Hobsbawm, *Labouring Men,* London, 1964, pp. 23–33.
90. Ibid.
91. Aled Jones, 'Garden of Eden: Welsh Missionaries in British India', in

G.H. Jenkins and R.R. Davies, eds., *From Medieval to Modern Wales: Historical Essays in Honour of Kenneth O. Morgan and Ralph A. Griffith*, Cardiff: University of Wales Press, 2004, pp. 264–80.

92. Lalsawma, op. cit., p. 5.
93. Ibid., p. 13.
94. Glyn Penrhyn Jones, *Newyn a Haint Yng Nghymru* (Famines and Epidemics in Wales), in Welsh cited in Lalsawma, op. cit., p. 14.
95. Ibid.
96. Mable Bickerstaff, *Something Wonderful Happened,* cited in Lalsawma, op. cit.
97. Lalsawma, op. cit., p. 15.
98. Jones, op. cit., p. 70.
99. Ibid., p. 73.
100. Llyod, op. cit., p. 162.
101. Lalsawma, op. cit., p. 152.
102. Ibid., pp. 152–4.
103. *Annual Report of Lushai Hills, 1919–20*, in K. Thanzauva (comp.), *Report of the Foreign Missions of the Presbyterian Church of Wales on Mizoram 1894–1957*, Aizawl: Synod, 1997, p. 63.
104. McCall, op. cit., pp. 196–7.
105. Zawla cited in Lalsawma, op. cit., p. 45.
106. Lalsawma, op. cit.
107. Ibid.
108. Ibid.
109. Saithanga quoted in ibid., p. 48.
110. Llyod, op. cit., p. 55.
111. Lalsawma, op. cit., p. 51.
112. Llyod, op. cit., p. 55.
113. Frederick S. Downs, 'Christianity and Social Change', in his *Essays on Christianity in North East India*, ed. Milton S. Sangma and Daivd R. Syiemlieh, Delhi: Indus, 1994, p. 224.
114. Frederick S. Downs, 'Christianity and Cultural Change in North East India', in his *Essays on Christianity in North East India,* ed. Milton and Syiemlieh, Delhi: Indus, 1994, pp. 184-97.
115. Susan Bayly, *Saints, Goddess and Kings: Christians and Muslims in South Indian Society 1700–1900*, Cambridge, 1989. Bayly's work refers to the situation in southern part of India but the situation was equally valid in north-east India.
116. Lalsawma, op. cit., p. 154.

117. Rev. J. Lorrain to Superintendent, Lushai Hills, 17 January 1925, DC (A) No. 16/24–5, 1924–5, MSA.
118. Ibid.
119. J. Needham SDO to Superintendent, Lushai Hills, 5 February 1925.
120. Report of Col. Hodgson, Director, Pasteur Institute attached to the letter to N.E. Parry, Superintendent Lushai Hills, from J.E. Webster, 19 March 1925.
121. Webster to Parry, 19 March 1925.
122. Ibid.
123. *Annual Report of Lushai Hills, 1919–20*, in K. Thanzauva (comp), op. cit., p. 63.
124. S. Bhattacharjee, 'Man, Calamity, History: The History of Earthquakes and its impact on North East India, 1226–1960', unpublished Ph.D. thesis, Assam University, Silchar, 2004.
125. B. David Raju, 'The Relief Activities of the Christian Missionaries in South Coastal Andhra During the Famine of 1876–78 and Mass Conversions', paper presented and abstract published in the *Proceedings of the Indian History Congress*, 53rd Session, Warrangal, 1992–3, Delhi, 1993. Abstract no. 96.
126. *Report of Sylhet Cachar 1929, Silchar District*, in Rev. Vanlalchhunga (compiled), *Report of the Foreign Mission of the Presbyterian Church of Wales on Sylhet-Bangladesh and Cachar, India, 1886–1955*, Silchar: Shalom Publishers, 2003, p. 297.
127. Reproduced in *Annual Report of the Baptist Missionary Society, 1901–30*, Aizawl: Baptist Mission Society, 1993, pp. 264-5.
128. Ibid., p. 267.
129. *The Life and Witness* as in note 34, pp. 62–3.
130. Lloyd, op. cit., pp. 65–6.
131. Ibid., p. 56.
132. Ibid., pp. 93-4.
133. Ibid.
134. Jones, op. cit., pp. 94–5.
135. Llyod, op. cit., p. 56.
136. Stokes, op. cit., pp. 1–80.
137. Ibid., pp. 30–1.
138. McCall, op. cit., p. 204.
139. Ibid., p. 204.
140. Richard Eaton found similar reason behind the Naga acceptance of Christianity *en masse*. The pre-Christian faith of the Naga, another

important tribe of north-east India, was such that Christianity could easily be superimposed on it as the Pantheon was transposed with Christ. Richard M. Eaton, 'Conversion to Christianity Among the Nagas, 1861-1971', *Indian Economic and Social History Review,* vol. 11, no. 1, 1984, pp. 1–43.

141. Stokes, op. cit., pp. 30–1.
142. Ibid., p. 40.
143. Ibid.
144. McCall, op. cit., p. 205.
145. Sajal Nag, *India and North East India: Mind, Politics and the Process of Integration, 1946–50*, Delhi: Regency, 1998, pp. 7–19.
146. McCall, op. cit., p. 199.
147. Ibid., p. 205.
148. Downs, op. cit.; and S.K. Chaube, *Hill Politics in North-East India,* Delhi: Orient Longman, 1973, also agree with this position.

CHAPTER 5

Famine as a Site for Politics of Nationalism

The next famine to stalk the hills of Mizoram was in 1959. But by then momentous changes had taken place in these hills. The British had withdrawn from the subcontinent. India had become independent, albeit partitioned. For the Mizo's it was not the first partition. They had already experienced an informal one in 1937 when Burma was separated from the British Indian Empire, which divided the kinsmen between two political areas. With the Partition of 1947 a number of them were marooned in the Chittagong Hill Tracts in present-day Bangladesh. During the past famines a good many of them had settled in the Jampui Hills of Tripura. This section also remained cut-off from their brethren as Tripura, a princely state, reverted to its old sovereign status. (It merged with India in 1949.) Thus, with the disappearance of the empire, all these kinship groups were artificially separated from each other. Political boundaries divided cultural communities and halted their natural exchanges. Moreover, in 1947, the Mizos were not the same people whom the British had once conquered. From a number of loose and mutually hostile tribes inhabiting the same hills, they had now forged a generic identity called the Mizos. With this identity they competed with the Indian national identity and launched a movement for self-determination. They dreamt of a destiny outside India as a separate, sovereign Mizo nation state. Alternately they could merge with their kinsmen in Burma. The Mizo's opted to join the Indian union since it held the promise of the egalitarian society and political structure. At that point, they also wanted to be free of the tyranny of the institutions of chieftainship. The genesis of Mizo nationhood could be traced in this struggle. The post-Second World War political situation unified the Mizos as a community. They formed a political platform called Mizo Union (1946), and fought on two fronts—against their marginalization in the Indian polity and oppression of the chiefs.[1]

However the decision to be absorbed within India was not unanimous. A small group comprising the chiefs and their loyalists opposed the move. This disgruntled section in time increased in number and covered the entire population, as the experience within Indian administration was turning disappointing. For example, in the Bordoloi Commission (1946) formed for representing tribal interests the Mizos were not given full membership as promised. Chieftainship was not abolished promptly as promised creating a great chasm within the Mizo society. A District Council was given to the people to satisfy their desire of autonomy but it was without any financial power. The Council was at the mercy of the Assam administration. To top it all, the Assam government tried to impose the Assamese language (The Assam Language Act 1960) while the Government of India attempt to introduce Hindi (The Official Language Act, Govt. of India 1960) on these Tibeto–Burman speaking people. The series of disappointments, breach of promises by the Indian state and confirmation of apprehensions turned the disgruntlement into a political movement. The Mizo movement for self-determination of the pre-Independence period was now renewed and a secessionist movement centring round Mizo nationalism emerged. The event, which facilitated this nationalist politics, was the famine of 1959. It was around this famine that the Mizos constructed the theories of Indian neglect and betrayal, mobilized people behind the ideology of Mizo nationalism and started the Mizo movement for secession. To understand the development of the politics of nationalism, which had been generated around the famine of 1959, an understanding of the pre-colonial developments in the hills of Mizoram is essential.

The Dilemma—to Merge or not to Merge

The British decision to withdraw from its Indian Empire in 1946 came as a surprise for most of the tribes of north-east India including the Lushais. They were complacent that British rule was permanent in the Mizo Hills. But the British decision compelled them to think about their future as they did not want to be governed by an Indian administration without having any say in the transfer of power. For 'having a say' they needed to politically activate and organize themselves. Under the leadership of R. Vanlawma—the first matriculate among the Mizos, who had the experience of organizing the Young Lushai Association, the Mizo Union—the first political party was launched. Originally this party was called Mizo Commoner's Union as the party's main objective was to get rid of the oppressive chiefly rule. But it was changed to Mizo Union so that the division in Mizo society was not reflected in it. However the character of the Mizo Union did not change with the nomenclature. It remained an anti-chief party. Although legally the withdrawal of the British from the Indian Empire would automatically entitle the new Indian regime to inherit the entire British India except those areas which were effected by Partition, the Indian National Congress which was leading the Indian struggle for freedom very courageously upheld the Principle of Rights to Self-Determination of Nationalities. Its resolution in 1946 that 'it cannot think in terms of compelling people in any territorial unit to remain in an Indian Union against their declared and established will'[2] encouraged ethnic groups to think in terms separation and even secession. This included the Mizos and the Nagas. Although these areas were 'Excluded Areas' under the Government of India Act of 1935, they were still an integral part of British India. They were only excluded from the political processes that were introduced by the Act. But due to the apprehension of the tribals about their position under Indian rule, these tribes too started to think of a possibility for a separate existence.

The Mizo Union's leadership was inclined to merge with India, as it would entail the abolition of chieftainship, which they so desperately desired. The Indian National Congress was committed to the abolition of such medieval institutions like landlordism, chieftainship and so on. This was the main attraction of the Mizo Union towards India. But the same promise alerted the powerful chiefs. They grouped together under another banner called the United Mizo Freedom Organization, which stood for the independence and sovereignty of the Mizos. Even the Mizo Union party had split into two: left and right wing factions. The right wing was dominated by the chiefs, which opposed the merger with India. The debate between the two groups reflected the dilemma in the hills. In a public meeting held in Aizawl, headquarters of Mizo Hills district to review the situation in 1946, the Mizo leader Vanlawma stated the reasons why the Mizos should refuse to be join India and declare independence:

In the ancient past Mizoram was not under anybody's government. Now that the British who controlled us are about to leave the Asian subcontinent we should resume the status we held before the arrival of the British. We should demand total independence. . . .[3]

The Mizos attending the meeting were greatly agitated. Some were for independence while others were fearful of the consequences. There was a public debate on the issue. Replying to the question as to why the Mizo Union had not aimed at independence right from its inception and the issue was being raised now, Vanlawma said,

When we formed Mizo Union party the British administration was not clear as to when and how they were going to leave India. Under them the country was taken care of nicely and if we had mentioned independence when we started the Mizo Union party, the British would not have let us start it at all. But now that India is going to obtain independence, we feel that they [Indians] will be ruling our country and not considering our own interests. However the attitude of the Indian people is becoming clearer. They failed to carry out their promise to us: that we would have full membership on the planning board and have asked us to be co-opted members only and might intend to give us still less than self-determination in the future. Now that we know that they are not going to carry out their promises our future looks very uncertain. Therefore we must govern ourselves. At the moment we have enough supplies and if we lack supplies we will still find some other country to help us. And if we look at our natural resources and increase our produce by improving our farming system we will be able to produce a sufficiency of things. Now is the time to fight for our independence.[4]

Vanthuama the new General Secretary of MU countered:

It is impossible for us to fight for our independence now. If we look around us, we see the Darwin theory—the more powerful swallowing up the less powerful. If and when we are truly more powerful, we will swallow the Indians and if they are more powerful than us, then they will swallow us. Besides if we are independent where will we get salt and iron ore to make our farming equipment and how are we going to make money.[5]

There was a sharp difference of opinion polarizing the Mizos into opposing groups. Vanlawma and Vanthuama were the leaders of these opposing camps. Vanlawma replied to Vanthuama's argument saying:

Pu (Mr.) Vanthuama's statement on Darwin's theory seems to me to be an attempt to escape reality. We all know for sure that we, the Mizos, are much smaller and less powerful than the Indians. For that very reason we created the Mizo Union Party. . . .

Concerning salt and iron our ancestors, though less advanced than we, were self-sufficient and even made their own guns. If our ancestors knew how to trade with their neighbours we certainly ought to be able to take care of our own affairs. Concerning money we can use it as the rest of the world does. If we have enough food there is no need in fact to be unduly alarmed about our future.[6]

The visit of the Bordoloi Committee witnessed the intensification of the separatist movement. The left wing Mizo Union whom the committee had accepted as the mouthpiece of the Mizos joined the Bordoloi Committee and expressed the desire to be integrated into the Indian union. This gave the accession a seal of finality. It upset the section opposed to integration. Now a concerted effort was started in favour of independence. Pachunga, the president of the right wing Mizo Union along with Dahrawka and Hmartawnpunga two other prominent Mizo leaders published a political pamphlet, which read:

We Mizos have nothing in common with the *Vai* (Indians). If we commit ourselves under the Indian government, we will be swallowed by the Indians because they are greater in number than the Mizo people. Until the British came, we Mizos had nothing to do with the *Vai,* now that the

British are leaving we should get out of the British government to be as we were before, namely, free. The Mizos are neither slaves nor possessions; therefore we should not allow ourselves to be treated as such, having to change owners. The Mizos should stand firm together and defend Mizoram for the Mizo people. Whatever may come Mizoram is for the Mizos.[7]

Earlier than this (5 May 1947) a pamphlet entitled *Zoram Independent* written by one D. Ronghaka was published and distributed in Aizawl. The pamphlet advocated that the Mizos should declare themselves Independent. The English translation of the Mizo pamphlet reads:[8]

Every nation in the world strives for independence. India had struggled long to secure its independence. So have the Muslims of India for their independence. If the Mizos do not fight for their independence, they will remain slaves (*Tuk Luh Bawi*) which practise has been abolished long ago. We should fight for independence to avoid becoming slaves again. The fact that we speak one language (which proves that we are one people) is reason enough for us to strive for independence. Some of us are of the opinion that we should remain within India for the present and seek independence later. But this would prove very difficult. In time there would be some other pharaoh who had power and did not remember our Joseph and if he ordered you to kill your first son you would not be able to disobey it if you are part of India. Our succeeding generations would be in a difficult situation. If we are independent, all of us will be happy because then we will be working for our own future. It might be difficult at the initial period but it would be a worthwhile struggle.

We have heard people say that Mizoram is a rich country. We have delicious evergreen trees, bamboos, vines and many other plants from which medicines and herbs can be made. We have plenty of trees bearing fruits. We can increase the number of such trees and sell the fruits to our neighbouring countries. We will also increase our handicraft. We will improve our education. We will have the power to make machines. While the British government ruled us, we could not even make a gun because they provided it to us. We can make anything we want, if we are independent.

Because of our religion alone we should be away from the Indians. All around us, different religious groups seem to form their own countries. The Burmese are Buddhists. The Indians are Hindus and the Pakistanis are Muslims. Why should not the Mizos who are Christians have our own sovereign country? As Prophet Isaiah said, 'Have faith and it will be done.'

Let us say, if these people can have separate country, we Mizos can have one too. This is the time for us to take our religion seriously. If Prophet Isaiah had been alive he would have said, 'Believe in God and start your own government'. At the moment some of us are concerned about their own selfish interests. But let us look at the future not just our present. Let us strive to keep this country for our own children and grandchildren, not just think about ourselves.

Following this, a plethora of pamphlets were distributed in Mizoram through which debate took place across Mizoram on whether Mizos should remain within India, join Burma, remain under the British or declare themselves independent. Thus the more closer Indian Independence came, the more polarized the Mizo leaders became over the issue of joining India or remaining independent. Vanthuama led the pro-India group. Pachunga, who was stripped of his Mizo Union Presidentship, led the anti-merger group. There was also a debate over the issue of remaining a part of Assam or be governed by the Indian government directly in case the Mizo Hills merged with India. In the midst of all these both the parties seemed to agree with the point raised by Reverend Zairema and Reverend Thanlira, editors of *Mizo Daily* that Mizos should demand as much autonomy as possible even if a merger was agreed upon. H.K. Bawichuaka a leader of the Mizo Union demanded that the Mizos should be given adequate representation in the Assam Provisional Legislature, widest possible autonomy and amalgamation of all the Mizo inhabited areas into one unit. One of the members of Pachunga faction wanted the Mizos to join the Mizo areas of Burma and form a separate province therein.

The Mizo Union faction led by Vanlawma and Pachunga continued their campaign for an independent Mizoram. Vanlawma wrote a persuasive piece called *Khawi Lamah Nge I Kal Dawn*? *Quo Vadis*? (where are you going).[9] He refuted the argument that the Mizos were 'too stupid' to govern themselves. He provided instances from the Bible saying that like the philosophers of Israel who were afraid to leave Egypt but were guided by a fire till their destination was reached, the Mizos were saved from Japanese invasion. Similarly this time too God would certainly save them from the idol-worshippers (Hindus). Therefore Mizos should not be afraid of seeking independence.

Pachunga too wanted the British not to desert the Mizos until they were able to govern themselves. He sent a memorandum to the Constituent Assembly, the British Prime Minister Clement Attlee and the opposition leader, Winston Churchill pleading that the British should not leave the Christian Mizos to the mercy of the Hindu Indians.[10] Another similar pamphlet was circulated by K. Zawla, which lamented the Mizo's prospective merger with India and advocated Mizo independence.[11]

The pamphlet read:

It is most beneficial for the Mizos to have the British remain in Mizoram. Instead of having a limited, autonomous district attached to Indian government or Burma government it would be preferable to be a limited state attached to the British. Before it is too late, let us consider carefully what is best for us. Imagine the state of things in the near future with Hindi as the official language, because virtually no Mizo spoke Hindi. If you allow the British to leave now, you will greatly regret it later . . . to acquire the desired Mizoram and to make Mizoram a place in which Jesus could be happy to live let us ask God for guidance.

When I would go to have a quiet moment alone, not knowing exactly why, I would cry for all the Mizo men in the Mizoram villages because they were forced to carry baggage for the Indian and the British officials, but now I cried for the Mizos who had formal education, because they wanted to sell Mizoram to the Indian government. As soon as the Mizos joined the Indian government, they would have to pay such taxes as poll tax, land tax, home ownership tax, vehicle tax, bullock tax, car tax and school tax. They will pay more taxes than they ever paid to the British government. In effect, we will be slaves to the Indians.

Therefore, the best thing we can do is to remain under the rule of the British government here in Mizoram. If the British stay in Mizoram for three to five years in which time we learn from the British how to run our own government, this will be a happy occurrence. If the British remain in Mizoram, our chiefs will be ruling under some constitutions, and they will not be too much of a burden for the common people; and the superintendents, knowing the British occupation is limited will not become corrupted by an abundance of power. After 3 to 5 years, if we feel that we want to join either Burma or India, we can join them. We can always be the slaves of the Indians.

As a culmination of this process on 5 July 1947, another party known as the United Mizo Freedom Organization (UMFO) surfaced under the leadership of Lalbiakthanga an ex-Burmese Mizo military officer. Lalbiakthanga became the President of the organization while Hmingliana and L.H. Liana were Vice-President and General Secretary respectively. Lalbiakthanga believed that it would be difficult to talk of an independent Mizoram after 15 August 1947. Therefore complete independence and the unification of all the contiguous Mizo areas should be sought before the boundaries between India, Pakistan and Burma were demarcated.[12] Apart from the right wingers, the richer Mizo and the chiefs' council supported the party as well.[13] Its secessionist character and objective as evident from the name itself. It earned the support of the chiefs and rich businessmen who provided it with financial assistance. The chiefs' council selected Lalmawia, the founder leader of the Lushai Students' Association and an ex-army officer in Burma to work out a modified plan to keep Mizoram out of India and also mobilize public opinion in favour of it.[14] (Lalmawia later became the President of the party when Lalbiakthanga left the party.) But when the party prepared its constitution it did not make its secessionist character very explicit. This was done to depict that this party's aim was not different from the left wing Mizo Union party which was already immensely popular among the masses.[15] The constitution listed the objective of the party as:

1. To look for a country with which we could identify ourselves and to which we admit and which can provide us the benefits.
2. To promote true democratic spirits where people can choose their own leaders and to reject all traces of authoritarianism.
3. To make our country as strong as possible so that it need not depend on other countries.
4. To improve our culture.
5. To try to better out communication and understanding between ruler and the ruled.
6. To develop the best ways of self-administration.
7. To ensure freedom of speech and press.
8. To ensure freedom of religion.[16]

Lalmawia who served in Burma had contacts with the Mizo chiefs of Burma. It was also reported that on the instructions of the chief's council, he discussed the feasibility of forming a union of the hill areas of Burma and Lushai Hills with the tribal chiefs of Falam–Zahrelian region as well so that they could form a strong political-bloc. Lalmawia and the chief agreed to unite. But the Burmese government did not encourage the idea of merging Mizoram with Burma as it had already opted for India.[17] This fizzled out the UMFO movement for merger with Burma and with it the UMFO Party because by then India had become an independent nation.

One of the most important contributions of the UMFO was to officially start a movement for joining Burma. For some time the movement caught the fancy of certain sections of the Mizo people. The arguments advanced in favour of merger with Burma were that the Mizo tribal was not only ethnically closer to the Burmese but their languages were very similar. Economically, as a part of Burma, the Mizos could enter the international market for its bamboo and other goods. Politically it would be more advantageous to join Burma as it was smaller than India and hence Burma would grant the Mizos adequate representation in its political affairs, which was not likely in India.[18]

There was very little time for the UMFO to become popular or mobilize public opinion in support of its programmes. It could not make any inroads into the Mizo Union stronghold. The base of the Mizo Union was the masses that refused to receive the ideas of UMFO. The people wanted to be rid of the chiefs whereas the UMFO was the party of the chiefs working for the perpetuation of its rule. In fact, the party was known as the *Zalen Pawl*[19] or the 'Party of the Privileged'. Hence they rejected the UMFO and without the support of the common people the UMFO simply faded out.

Initially however, against the increasing momentum of the secessionist movement the Mizo Union seemed to be losing out. To gain losing ground they prepared to send its volunteers to the villages and interior areas to campaign for the merger of Mizoram with India amd familiarize the people on the issue of Indian

Independence. Vankhama and Lalrinliana were sent to Sialsuk (east of Aizawl); Vanlalliana (son of Pachunga) and Challeta were sent to Champai (east of Aizawl), Ngura and his friends to the north-east of Mizo Hills and Thangridema and Thantuma were sent to Baktawng.[20] These volunteers, during their extensive tour in the interior of the Mizo Hills discovered that the common people had no idea of the impending British withdrawal. Reeling under the oppression of the chiefs these villagers did not care much about the future political identity of Mizoram. What was more pressing was ridding themsleves of the oppressive rule of the chiefs. A folk song of the Mizos of that period reflected their desire:

bai thak arva
artui khawn leh
lal hnungzui reng ka ning tawh
kawltu chawi lai daltu an ni
sazai lian pui an ni

We are fed up submitting to the chief's orders. He wants us to do all his work. He demands eggs and chickens. We have to always carry out his orders and in the process get delayed for our own work. This is indeed a severe punishment.

Another similar song read:

zalen muhil lo tho r'u
ni a chhuak sang tawh eme
in kawmawl ahho reng hi
a lum lua e ka ti

Far in the horizon, the sun is making its appearance; it is the sun of freedom. As it rise high in the sky we all can feel the growing heat around us.

The Mizo volunteers reported this to the party from which the MU think-tank picked up the thread. It referred to the Congress's promise of abolishing of all forms of monarchy, land-lordism and chiefly rule and obtained a promise from the Gopinath Bordoloi headed Government of Assam, the representative of the Indian National Congress in north-east India, that after British withdrawal, chieftainship would be abolished. This hope spread among the masses. This created a sensational turn in Mizo politics. Enthused by the promise, people rallied round the Mizo Union. They organized a crusade against the chiefs and even launched a civil disobedience movement against them. As the tempo rose there was confrontation between the chiefs and the commoners. Alarmed at the prospect, the chiefs began to support the new party, UMFO, which weakened the Mizo independence movement and strengthened the merger movement.

The mood of the people was also recorded in a Mizo folk song:

India zawm duh chu lal banna
Independence duh chu lal lalna

It meant that joining India would ensure the abolition of chieftainship, but Independence threatened its continuation.

And,

Union le Union a dang mange
Union vantlang kan tanrual lain
Dawrpuii Union ve che rual elna

Though both the parties are called Mizo Union, they are different from each other. While the Mizo Union is trying to unite the people, the Dawrpuii (Pachunga) faction speaks ill of others and causes disunity.

In the meantime D.A. Penn (1946–7) replaced A.R.H. McDonald (1943–5) as Superintendent of Lushai Hills. He was there for a very short time and was in turn replaced by L.L. Peters (1947–9). The extensive support for merger sidelined the UMFO and strengthened MU. Enthused by this Khawtinkhuma of MU prepared for a procession in Aizawl to celebrate India's Independence on 15 August 1947. The Pachunga–Vanlanwa faction tried to prevent the procession and wrote to Khawtinkhuna to that effect.[21] It also asked the District Superintendent not to hoist the Indian flag in Mizoram on that day. The situation headed for a violent confrontation. On the night of 14 August a MU meeting took place in Dawrpui Veng. When the situation became very tense the procession was postponed. The Superintendent called a meeting on the same day, which was attended by about 50 Mizos where a resolution on the issue of manager to the Indian Union was passed, which read:

1. Resolved that owing to the unexpected acceleration of the date of transfer of power by the British Government and as the Lushai's have not yet been definitely informed in detail as to what is the proposed future constitution and form of administration of the district and as section (7) sub-section (2) of Indian Independence Bill does not clarify the situation, it is accordingly thought necessary that his Excellency the Governor of Assam should kindly inform them in writing as to what these are to be, also whether the Lushais are at this stage allowed the option of joining any other Dominion, i.e. Pakistan or Burma.
   Resolved further that the Superintendent, Lushai Hills should communicate the above request of the Lushais to the advisor to His Excellency the Governor of Assam in order to clarify these points.
2. Resolved if the Lushais are to enter Indian Union their main demands are:
   (a) That the existing safeguards of customary laws and land tenure, etc. should be maintained.
   (b) That the Chin Hills Regulation 1896 and Bengal Eastern Frontier Regulation 1873 should be retained until such times as the Lushais themselves through parallel district authority declare that these can be abrogated.
3. That the Lushais will be allowed to opt out of the Indian Union when they wish to do so subject to a minimum period of ten years.

Although the pro-Independence faction was in a minority in the meeting, it was able to secure major concession because L.L. Peter, the English Superintendent, had patronized them. The concession came in the form of Clause III, which stated that the merger of Mizo Hills with India was only for ten years and the former had the option to withdraw from India after the specified period of ten years. Superintendent Peter had no authority from the Government of India or the Home Government in England to hold such meeting or effect such resolution. The meeting also did not have any sanction from the Indian National Congress.

Hence the Resolution passed in the meeting turned out to be just a piece of paper, which had no legitimacy. It only placated the anti-India faction and eased the impending tension. On 15 August 1947, however, there was no untoward incident when the Mizo Hills formally became a part of India.

The Crisis

The advent of Indian Independence and the integration of Mizo Hills with the union did not signify peace or stability. The debate and the mobilization of the people over the issue of merger had thoroughly polarized society into two opposite camps: the chiefs and commoners or in other words integrationists or secessionists. While the chiefs vehemently opposed the loss of power, privileges and position, the commoners rejoiced the prospective freedom. The imminence of a violent clash could not be ruled out. The abolition of chieftainship required legislation that in turn required time. The newly independent Indian union could not do so immediately. The local administration took advantage of this. The British administration from the very beginning used the institution of chiefs for governing the Mizos and as such encouraged the power of the chiefs. The new British Superintendent of the Lushai Hills, L.L. Peters, who continued to hold the post even after Independence, perpetuated the policy. He not only began to show open favour to the chiefs but also safeguarded them and proved to be corrupt. It was reported that he regularly took bribes in kind like fowls, eggs and fish. These bribes could be for land grants to build a house and similar issue. Only the chiefs and such rich families could offer such bribes. Under him, the administration remained pro-chief even though the Mizo Union fought for the abolition of the institution. This was greatly resented by the Mizo Union. 'It is a sad thing that when the Mizos had high hopes for ending their miseries under free India, the first government officer of free India emerged to be the most corrupt person among the local administrators of Lushai Hills.'[22] In October 1948, Peters issued a circular to the chiefs that any subject who misbehaved with them must be reported to the superintendent so that the culprit could be punished. The order became an instrument of oppression. Frequently the chiefs would report cases of misbehaviours against commoners, mainly Mizo Union members, who would then be arrested.[23]

Pushed to the edge the Mizo Union leaders held a meeting in March 1948. The meeting demanded the withdrawal of Peters from the Mizo Hills failing which they would resort to non-cooperation with the chiefs as well as the administration. Accordingly a resolution was passed by the Mizo Union party to seek the removal of the suprintendent. The resoltuion stated, 'in order to bring the partiality of the administration for endurance and justice and bring back the rule of right in Mizoram it was decided to seek the removal of Superintendent L.L. Peters from Mizoram. If it is not effected immediately, the Mizo Union would resort to non-violent non-cooperation movement with the administration.' The Union fixed 27 December 1948 as the deadline. It also submitted a memorandum to the Government of Assam on 24 November 1948 detailing the grievance and complaints of the people against the Superintendent. A union delegation also met N.K. Rustomji, Advisor to the Governor of Assam and apprised him of the situation. The district administration reacted by arresting the Union General Secretary, M.Vanthuama and Treasurer, Lalbuaia. The President of the party, R. Thanlirah was initially arrested in Shillong and on his return to Aizawl, was arrested again. Several other prominent union leaders were also arrested and the press that printed the Mizo Union mouthpiece, *Thupuan* was closed. On 28 December after the deadline, the Union organized a mass procession in Aizawl marking the beginning of non-cooperation. Accordingly people defied the orders of the superintendent or his officials. The commoners refused to pay the customary taxes, like *sachchian*, *buhcchun* to the chief and refused to build his house as was required by tradition. It also refused to render coolie services to the administration. There were widespread attacks on the houses of chiefs, *upas*, and *ramhuals* and their gardens destroyed. The administration responded by a massive crackdown on the activists, large-scale arrests and fines were imposed. When Rustomji, Advisor to Governor of Assam reached the hills for an enquiry he found 'the atmosphere disturbingly tense. . . . The chiefs of course urged the most drastic measures against the Mizo Union leaders a considerable number of whom had already been put up in the lockup by the overzealous Superintendent. The chiefs complained that the Mizo Union was misleading the public. They were storming the houses of the chiefs and threatening murder. The administration was on the verge of collapse. . . .'[24] Rustomji's intervention calmed nerves and the movement was called off by the Union. The administration released the arrested leaders. However the public opposition to the chiefs did not ease. It did only after the abolition of chieftainship in 1954.

Social Stratification in the Tribal Structure

Among the much stratified Mizo society the primary difference was between the chiefs and the commoners, which was politicized and thus consolidated during 1946–54. The Mizo villages were ruled by hereditary chiefs largely belonging to the Sailo lineage of Thangur clan of the Lushai tribe. In spite of the fact that the Lushai practised ultimogeniture, the chieftainship used to be inherited by the eldest son. Each chief (*lal*) had his own territory defined by the British after annexation where he was supreme. He was the leader in war and administration in village. The chiefs allotted plots of land to people. But people hardly had time to farm their lands, as they had to cultivate lands belonging to the chief. Every household had to pay a *fathang* of Rs. 2 or two tins of paddy to the chief, one foreleg of the animal killed as *sachchian*. The British legitimized the existing power and privileges of the chiefs after it took over the Mizo Hills. Sanctioned by the colonial administration, the chiefs emerged more autocratic. All laws and order were promulgated through the chiefs. There were about 60 chiefs when the British arrived. But the number continued to increase under colonial patronization and by 1955 there were more than 253 chiefs among a population of 1,50,000. In addition, there were another 50 chiefs in Pawi–Lakher region under the district council of the Lushai Hills and in 1956. The administration of the district was vested in the hands of Governor of Assam, the district executive administration was in the hands of the Superintendent, his assistants, the chiefs and headmen of the villages. The district was divided into two sub-divisions, which was further sub-divided in 1901–2 in the circles comprising a number of villages. The district was divided into 18 circles—12 in Aizawl and 6 in Lungleh sub-division with an interpreter in-charge of each. Those circle interpreters toured their respective circles and all orders for the chief and people were sent through them. The cardinal principle was to administer the district through the chief. The staff of the circle formed the intelligence link. On the other hand the commoners had to approach the administrators through their respective chiefs. Occasionally the chiefs held *durbars* to communicate their idea and declare their loyalty to the British administration. Although the British perpetuated the institution and increased their number, the power of the chiefs had begun to wane during the later part of their rule. In other words, covertly and gradually, the British administrators themselves were taking over the actual powers of the chiefs. The common people did not realize it as they continued to be controlled by the chiefs. They adjusted to it grudgingly and were dreaming of reverting to their pre-British autocratic days. In fact, the Mizo chiefs were all traditionally from the Sailo clan. The British intervened in this tradition by arbitrarily selecting chiefs from other clans as well. This explained the increase in their number from 60 to 250. This was done to counter the power of the Sailo chiefs and also to build a support base among the chiefs. The Sailo chiefs were aghast at this but could do nothing. Not only this, the administration also curbed the actual power of the chiefs. It was interesting to note that on the one hand, the colonial administration ruled the people through the institution of chiefs and on the other curbed the power of the same chiefs. The decline in the power of chiefs, in fact, had actually begun with the Land Settlement of 1898–9 by which the chief's power to grant land plots to commoners were abolished. Subsequently, to meet the exigencies of the political situation, the British government arbitrarily curtailed a number of their traditional rights, e.g.

1. Right to order capital punishment.
2. Right to seize food stores and property of villages who wished to transfer their allegiance.
3. Proprietary rights over lands now arbitrarily reserved by the government.
4. Right to tax traders doing business in chiefs area.
5. Right to freedom of action relating to making their sons the next chief.
6. Right to help *bawis.*
7. Right to claim property.

The Sailo chiefs never reconciled to the loss of power. They were hoping for the exit of the British from their *ram* (area of jurisdiction) and waited for an opportunity to regain their lost empire. There was restlessness among them throughout British rule which made the latter fear a rebellion any time. The administration referred to it frequently.[25] In fact, during the tenure of A.G. McCall, the superintendent ordered an enquiry into the causes of such restlessness.[26] The Mizo middle class—product of the new education and enlightenment—saw through this policy and were in favour of the abolition of this oppressive institution. The popular sentiment of the people was enlisted for mobilizing opinion and on the slogan of abolition of chieftainship and establishment of a new egalitarian society in Mizoram, the Mizo Union succeeded in neutralizing any pre-emptive move to opt out of India. In the process it antagonized the chiefs who did not want to stop its progressive decline. The non-cooperation movement against them betrayed the peoples' anger. The last blow was the Lushai Hills Reorganization of Chiefs Rights Act, 1954 by which the power and privilege of the chiefs were taken away abolishing this age-old Mizo institution. The Mizo Union fulfilled most of its promises. As a result it captured 23 out of 24 seats in the District Council election in 1952. The party secured the abolition of chieftainship in 1954 thereby reorganizing the rights and privileges of the 259 Lushai chiefs and 50 Pawi–Lakher chiefs by the

Assam government with effect from 1 April 1956 and 15 April 1956 respectively. In 1952 personal residential surcharge was also removed. In 1953, the system of imprest coolies or compulsory labour and traditional due to the chief too was abolished. These changes were perpetual defeats of the chiefs about which they could do nothing. They could not fight in the District Council as they had only one representative in the council, who belonged to the UMFO party. The decline and fall of the chiefs from grace was complete by recent legislations. They first unified and tried to resurface through two political parties—Mizo Union (right wing) and UMFO without much success. The angry chiefs were looking for an opportunity to strike back and re-establish a '*ram*' where they could re-establish their authority and supremacy. The 1959 famine provided that opportunity.

Coming of the Famine

As per the Mizo calculations, the next famine (after 1929) was to strike in 1959. But this time neither the 'paternalistic' British nor the powerful chiefs were there. A great number of foreign missionaries also had left the hills. The affairs of the Mizos were theoretically in their own hands, but practically they had to depend on the Government of Assam for almost everything as the district council had no money of its own. It was essentially a legislative body. The execution was under the Government of Assam. It meant the leadership had to come forward to organize the people to combat the menace. The past experiences of combating the famine were there. But the last famine occurred in 1929–30. The generation, which witnessed the famine relief under the British, were no longer in the scene and the others were too young to remember anything. The fear of the impending famine was there in the minds of the people immediately after Independence. This was evident from the fact that even as the tiny Mizo population living in the far away hills were ravaged by one crisis after another from the time of the withdrawal of the British and the whole society was polarized into two warring groups, a non-governmental organization called Anti-Famine Campaign

Organization was formed on 21 July 1951 under the chairmanship of D. Rokhuma. It had members from a wide range of professions. The Superintendent of the hills, subsequently redesignated Deputy Commissioner of Lushai Hills Autonomous District, S.N. Barkakati (1951–3), when asked to bless the organization also assured his services at all times for the cause.[27] The organization was later reorganized with the Lalmawia (MLA) as Chairman, H. Khuma as Vice-Chairman and Rokhuma as the Organizing Secretary. Pachunga was the Treasurer, Lalbiaka was Assistant Secretary and Vanthuama a member. The organization printed pamphlets and distributed them among people detailing the do's and don'ts about the impending famine. The AFO was the only non-political body at that time with a purely humanitarian objective. As a result it was able to draw people from rival organizations like the MU and UMFO. But AFO was not the only functioning organization. Vanlawma, sidelined after an unceremonious exit from Mizo Union, had formed the Mizo Cultural Society. Although the society was apolitical, most of its members were from the erstwhile Mizo Union right wing faction or the UMFO, both of which had favoured secession of Mizo Hills from India. This cautioned the Indian administration against them. The Deputy Commissioner, L.S. Ingty (1958–60) by an order announced that no government employees could join this organization. This resulted in the decline of the organization in its importance as it was seen as a banned party on the one hand and on the other hand a pro-chief secessionist organization.

At the turn of the year 1954 *rawte* (one variety of bamboo) started flowering in certain pockets of the country. The flowering was accompanied by an increase in rat population. By 1956 another kind of bamboo locally known as *phulrua* showed signs of dying again and by 1957 the signs of the impending *mautam* were visible. Bamboos began flowering and bearing fruits everywhere. There was simultaneous increase in the multitude of rats. The dread emerged as the previous two harvests were poor and the food production was enough to last only 9 months of the year for each household. The AFO was trying hard to counter the situation through peoples' participation. It encouraged the people to kill the rats and produce alternative food product like fruits such as banana, which were plenty in the Mizo Hills. This was similar to the ways the Mizos fought against the famines in pre-British days, which helped the people to live through the famines. But the Government of Assam, which replaced the British administration in the hills showed no signs of any pre-emptive measures. When the Supply Minister of Assam visited Aizawl on 25 May 1954, the AFO apprised him of the impending danger and requested him about the necessity of setting up of food stores as well as strengthening of the supply network as the British did in 1912 and 1929. To the utter surprise, the minister dismissed the whole idea of rats causing in a famine as tribal superstition. When insisted he casually assured the AFO activists that even if it occurs, the government would not leave the Mizos 'to die in hunger'.[28] In a strange action even the District Council banned the use of pesticides, which the AFO was using to kill the rats. The decision was however changed on the pressure of the AFO.[29]

Beginning of Politics

The hundred years long colonial rule in the hills had produced and prepared a modern middle class. While the chiefs refused to give up their traditional position even in their transformed social structure, the middle class wanted to reshape their respective societies according to modern ideas. This conflict between these traditional and modern leadership was evident in all the tribal societies in north-east Indian the Khasis, Nagas and—the Mizos.

The adoption of the Constitution by Independent India and the subsequent appointment of the State Reorganization Commission (SRC) generated immense enthusiasm and hope among the tribals of north-east India. Most of the tribals were not comfortable within the composite state of Assam. They felt squeezed between the Assamese and Bengalis who were mostly Hindus and Muslims, where the political hegemony remained with Hindu Assamese elites functioning within a casteist structure. Therefore at the first opportunity they demanded autonomy in the form of a statehood exclusively for the tribals. The demand was voiced first by the Assam Hill Tribal Leaders Conference to the SRC, which was promptly rejected by the commission. When Jawaharlal Nehru visited the region on 28 August 1955, a delegation of hill leaders demanded a hill state on the ground that the tribal areas remained backward as the thrust of Assam's development plan was the plains. They also resented the prospective imposition of Assamese language on them. Assamese language was actually imposed on them later through the promulgation of Assam Language Act 1960. From 26 to 28 October 1955, the hill leaders held a conference in Aizawl where a decision was taken to merge all the tribal political parties to form Eastern India Tribal Union (EITU). The most powerful political party of the Mizo Hills—Mizo Union, however, preferred to maintain its separate identity. But during the second election to the District Council, the Mizo Union found that it had lost some ground. The reasons were not far to seek. After the abolition of chieftainship, the authority of distribution of urban lands was vested with the District Council, which was dominated by Mizo Union members. There was widespread allegations of favouritism in the distribution of land against the Mizo Union members. Meanwhile Bimala Prasad Chaliha replaced Bishnuram Medhi as a new Chief Minister of Assam. Although Bishnuram Medhi's tenure was not remarkable for anything, he had continued the benevolent policy of his predecessor Gopinath Bordoloi as far as the tribes were concerned. The north-eastern region had a huge tribal population of immense diversity who were in a very delicate stage of their history. The first Prime Minister of the country, Jawaharlal Nehru's sympathetic concern for these tribes had helped shape post-colonial India's tribal policy. The historical record of the majority Hindus in dealing with tribes was one of incessant persecution. In north-east India Gopinath Bordoloi put Nehru's dreams for the tribal people into place. Riding against the staunch opposition of his Assamese colleagues in granting autonomy to the tribals Bordoloi showed extraordinary courage. His contributions to tribal welfare had not been fully appreciated. He was the architect of the Sixth Schedule of the Indian Constitution, which subsequently became the lifeline of the tribals in north-east India. But Chaliha, the new

Chief Minister of Assam was not of the same disposition. The violent turn to the Naga secessionist movement had made many statesmen in the region rethink tribal policies of the government. The separatism and autonomy movements of other tribal leaders threatened a balkanization of Assam. There was an increasing insecurity among the Assamese regarding their tribal neighbours. Not only their political hegemony but also the territorial integrity of the Assam state was in danger. Chaliha worked against this backdrop. The constant complaints of the Mizo Union regarding the government's indifference towards the Mizos made him dislike the Union. He was contemptuous of the popularity of the Mizo Union Party. He had his opportunity to strike when the Mizo Union lost two seats in the election of 1957. The loss of two seats created an impression on Chaliha that the Mizo Union was a spent force. Although the Mizo Union was a consistent supporter of the Congress, Chaliha opted for a member of the UMFO, an avowed secessionist party, as the parliamentary secretary of his government. The Mizo Union representative was given only the position of chief parliamentary secretary. Although in India a higher position is pitied MU against UMFO. The Mizo Union was displeased with this inadequate reward. Even for the post of Minister-in-Charge of Tribal Areas Development, the Chief Minister chose Williamson Sangma of EITU in preference to a Congressman. The policy of deliberately sidelining the Mizo Union by Chaliha created bitterness between the Assam government and the Mizo Union. Chaliha actually offended the Mizo Union by declaring its members as corrupt. The estrangement between the Chief Minister and the Mizo Union reached its peak on the eve of the great famine of 1959.

As the famine became imminent, the Mizo Union, which was the ruling party in the District Council, began to take steps to pre-empt the hardship that the impending famine was likely to cause to the people. On 29 October 1958, the council passed a unanimous resolution.

With the flowering of bamboos in the Mizo district, rat population has enormously increased and it is feared that in the next year (1959) the whole District would be affected. As a precautionary measure against the imminence of famine, following the flowering of bamboos, the District Council feels that the government be moved to sanction to the Mizo District Council a sum of Rupees 15 lakhs to be expended on a test relief measures for the whole of the Mizo District including the Pawi–Lakher region.

The resolution was then forwarded to the Assam government. But the Chaliha government ignored the demands of the district council. This looked deliberate mainly because Chaliha had developed a tremendous dislike for the Mizo Union, which commanded the District Council.[30] Besides people at the helm of Assamese affairs were not familiar with the history of the Mizo Hills. They did not believe in the reported connection between bamboo flowering and increase in rat population and the consequent rise of famine. Therefore the government neither paid any heed to the resolution of the District Council nor did it take any other precautionary measures. The AFO, the most prominent non-govermental organization with no political affiliation was making serious efforts. Its organizing secretary, Rokhuma began to submit fortnightly reports of the crops situation to the central government along with a monitoring report on the increase in the rat population since 31 October 1957. The AFOs request to take immediate action moved the central government into action. It sent a plant protection officer (in-charge, north-eastern region), Sundharan Pillai to Mizoram to study the phenomenon of bamboo flowering. With the help of Rokhuma, Pillai recommended the opening of an office of plant protection in Aizawl itself, which was set-up on 28 May 1958. It appointed a local tribal officer as the plant protection officer. The District Council too vested the responsibility of educating the villagers about the procurement and use of rat-poison with the AFO, which it did effectively through its huge volunteer force working in the interiors.

The much-feared *mautam* meanwhile happened much earlier than expected. But contrary to the usual pattern, it affected the Mizo Hills in a phased manner. In 1958, famine was felt in areas lying to the east of Turini River. But by 1959 it had affected the rest of the Mizo Hills.

Operation Famine Relief

The famine caught the Assam government completely unaware. In fact, the famine had began in 1958. But by 1959 the government had still not officially recognized it as famine. Then, relief could not be sanctioned under the government rules. While the government slept over the files, rats were busy devouring foodstuff. It destroyed rice stocks virtually overnight. In one place rodents destroyed two acres of standing crops in one night. Although it had been anticipated for many months, the famine came suddenly and completely, and before people realized, the food stock was exhausted. In Pawi–Lakher region there were quite a few starvation deaths. It was reported that Reang and Chakma villagers were surviving on wild *arul* (a kind of grass) in the absence of any other food. It was around this time, on 19 January 1959, the finance commissioner from the central goernment visited the Lushai Hills. He saw the ravages of the famine and accordingly recommended immediate supply of food crops to Mizoram along with 6 trucks and 150 jeeps for carrying the foodstuff to the famine affected areas. But the Assam government was late in reacting. The famine had begun in the last month of 1958. The relief began to be issued only from August 1959. It was only when, on the desperate appeal of the district administration, the Tribal Welfare Minister of Assam, Williamson Sangma visited Aizawl and a angry public demonstration confronted him that he recommended the declarations of the Mizo Hills as a famine-affected area. As a result of this declaration relief was sanctioned for the famine-affected people. Initially the Assam government under-estimated the severity of the disaster. So relief was slow to come in. The indifference and callousness of the Government of Assam led to a huge debate about the Assam government's attitude towards the plight of the Mizos in the District Council. The District Council charged the government as 'incapable'. One member of the District Council Vanlalbiaka contrasted the indifference of the Assamese leadership with the all out effort and benevolence of the British in times of such distress. He concluded saying 'If we continue to be neglected . . . the peoples' feeling will be for secession from Assam'.[31] As the government sensed the growing alienation of the Mizos it increased the magnitude of relief measures though by then the worst was over. In total, the government granted a sum of Rs. 190 lakh on famine relief. The relief was sanctioned under various categories (heads):

1. *Gratuitous relief*—This relief was meant for the poor without any food assistance. The beneficiaries were not required to repay the advances in cash. The government spent a sum of Rs. 4,91,664 on this head.
2. *General relief*—This category was meant for all famine affected people especially in the rural areas. The government spent Rs. 28,65,834 on this account.
3. *Subsidy*—In order to make rice and other foodstuff available at a reasonably low rate, the government highly subsidized the transportation cost. An amount of Rs. 56,74,125 was spent on this account.
4. *Medical aid facilities*—Famine was accompanied or followed by epidemics. The government sanction a sum of Rs. 6,31,655 to procure medicines and health care facilities for the affected people.
5. *Agricultural schemes*—The government also sanctioned Rs. 4,08,761 on the free distribution of seeds and various agricultural inputs, pesticides for the farmers to start cultivation for the next season.
6. *Agricultural loan*—The government also sanctioned a sum of Rs. 57,22,423 from which it distributed agricultural loans to the farmers for the purchase of seeds, paddy, etc.
7. *Purchase of vehicles*—The communication system was virtually non-existent in the Mizo Hills. Few jeepable roads were there, few more were constructed during the famine relief operation. Since heavy Trucks could not ply on these roads. Willy's jeeps were purchased and stationed at the deputy commissioner's office in Aizawl for use to the distant villages. On this head, a sum of Rs. 18,47,429 was spent.
8. *Air dropping*—Foodstuff had to be air lifted from Silchar and air dropped to places like Lunglei and Champai, which were not connected by road. Therefore rice was air lifted from Jorhat

and Calcutta quite a few times. The expenditure on this account came to Rs. 87,567.

9. *Godown construction*—In order to store large amount of food-supply in good condition, construction of godowns were required at strategic places. The government spent a sum of Rs. 62,479 for this purpose.

Thus a total sum of Rs. 17,79,481 was spent on the famine relief in the Aizawl district alone. In total 3,32,390 people were affected in the Mizo Hills and a total of Rs. 1 crore 90 lakh was spent. However, the delay in providing the relief had already done the damage.

Politics of Famine Relief

Initially the Assam government decided to distribute the relief material to the famine affected people directly. This was because it did not want to involve the Mizo Union in the distribution of relief. This was done to marginalize the party. In other words, it by-passed the District Council and used the opportunity to cover the ground it lost due to its delay in sanctioning relief and improve the image of the administrative machinery. When asked about non-involvement of the District Council in the relief operation, the Chief Minister charged the Mizo Union councillors with corruption.[32] The Deputy Commissioner's office was given the responsibility of distributing relief. The idea was to counter the growing perception among the Mizos that the Government of Assam was insensitive and callous to their sufferings.

The Mizo Union, which controlled the District Council saw through the government's game and realized the fatal result of its non-involvement in the relief operations. It had already received a jolt in the 1957 election. The Assam administration's policy could seal its fate in the next election. The fact was that the Mizo Union through the District Council was the first to seek the attention of the government at Shillong of the imminent famine and it had demanded a meagre sum of Rs. 15 lakh to take precautionary measures, which was rejected. Even after that it kept the government informed about the volatility of the situation. But now that more than 10 times of that amount was being spent on relief, the Mizo Union was being excluded from the operation.

However, when there were complaints that the relief was not reaching the needy and the affected, the Chaliha government accused the Mizo Union of 'non-cooperation'. The District Council claimed that they did whatever possible under their limited capacity. The government denied cases of starvation deaths reported by the council. The fact was relief could not be reached to deserving people mainly because the District Commissioner's office neither had the infrastructure nor the manpower to handle such a massive relief operation. Moreover the communication system or the road network was virtually non-existent. Places like Demagiri could be reached on foot only. Champhai was seven days walk from Aizawl and north Vanlaiphai four days walk from Champhai. The road to Aizawl and Lungleh were only jeepable and heavy vehicles could ply with enormous difficulties. These were only fair weather roads and during monsoon remained mostly cut-off. Most of the roads connecting villages began to be constructed only during the relief operation and this took time, as there was dearth of labour. Hence despatch of relief material was delayed. The border tribes like the Buddhist Chakmas and Hinduized Reangs complained of non-receipt of relief. Their complaint was that as non-Christians, they were being discriminated against by the Mizos who were Christians.

The Mizo Union made a strong demand that the relief be distributed through the District Council as it had the access as well as manpower to reach all areas of the hills including the interiors. A reluctant Chaliha government was looking for an alternative agency for channelizing the relief material, which could be a counter to the Mizo Union. With the blessings of Chaliha the almost defunct Mizo Cultural Society was converted into a famine relief non-governmental organization called the Mizo National Famine Front (MNFF).[33] Ignoring the Mizo Union or the AFO, the relief operation was handed over to the MNFF. In fact, 'the Chaliha government was more concerned with district level politics than the organization of relief work to the distressed people.[34] The

Mizo Cultural Society was formed by Vanlawma after his unceremonious exit from the Mizo Union, which almost ended his political career. John F. Manliana, a public works department contractor, kept the organization alive. After it became the MNFF, Laldenga joined it and breathed fresh life into it. Laldenga hailed from Papui village, 8 km north of Aizawl town. He passed Middle English Exam and served as a lower primary school teacher under the control of missionaries. During the Second World War, he joined the army and became a Havildar Clerk. During the pre-Independence days, he was a member of the Mizo Union right wing and an ardent believer in the secession of Mizoram. After the Independence of India, he joined the Mizo District Council as an accountant and to escape charges of misappropriation of District Council funds, he resigned from his post and joined the MNFF.[35] Laldenga was an excellent organizer and orator. He haboured hatred towards Indian plainsmen, which he combined with his faith in the sovereignty of Mizoram. All these came to excellent use during the famine relief of 1959–60. On joining MNFF, he immediately put both Manliana and Vanlawma into secondary place. Chaliha's patronization made him the *supremo* in the organization. Chaliha was monitoring his rise. In order to placate Laldenga who had been lately propagating the slogan of 'Mizoram for Mizos', the Government of Assam sought his assistance in famine relief operation, even though the district commissioner, the District Council office and the AFO were already in the thick of it. Thus 'relief turned to be a salient form of patronage in the famine stricken hills of Mizoram'.[36]

The MNFF was also successful in promoting its own agenda during the operation. It recruited Mizo youth and sent them to remote villages. The Mizo villagers only saw the MNFF volunteers delivering foodstuff in their villages and gave all the credit to Laldenga and all his band of volunteers. Even though it was the Assam government which sponsored the MNFF for the distributions of its relief material, the MNFF spread canards against the Chaliha goverment. Being in close contact with MNFF volunteers, the villagers believed that the Assam government had neglected the Mizos and had not taken adequate remedial measures before and during the famine period. As the situation improved, the MNFF under the leadership of Laldenga, Lalnunmawia, Saing Lella and Vanlalhruia got all the credit and praise for the supply of foodstuff to the remote villages. The bulk of the relief funds were spent by way of subsidy on transport of grain, purchase of vehicles and petrol and construction of godowns for storage of rice. Whatever little foodstuff reached the remote villages were the ones carried by people as headloads. In a hilly terrain it was difficult to carry more than 25 kg and that too a maximum distance of 15 km. Defective packing rendered it unsuitable for air-dropping which led to wastage. Due to bad weather and hilly terrain food meant for certain areas were air-dropped in other areas. As a result certain areas had excess supply while others did not receive any relief at all, although on paper they did. In the absence of proper supply of foodgrains from Silchar, the only plains outlet after Partition, people in the interior villages got the wrong impression that the plains people were deliberately holding back rice bags in Silchar. The MNFF propagated it as an economic blockade by the Assam government. When complaints were made, the Assam government blamed the Mizo Union, the Mizo Union blamed the Assam government and MNFF blamed both and bagged all the credit for providing succour to the Mizos in their time of greatest crisis.

The Balance Sheet: Facts of Relief

According to the census, the total population of Mizoram in 1961 was 2,66,063 of which urban population was only 14,257 and rural 2,51,806. It was a small population for which Rs. 190 lakh were sanctioned for famine mitigation. A sum of about Rs. 170 lakh were actually spent. Out of these only (4,91,064 + 28,65,834) 33,56,898 was spent on foodstuff. The total food grains supplied was 2,12,034 maunds and 10 seers, 2,85,748 maunds and 19 seers by air dropping and, 5,74,244 maunds, 12 seers by road. The rest of the money was spent on subsidy on the transport of grains, purchase of vehicles, cost of petrol, cost of air dropping and construction of rice stores. The import of the food grains was sufficient for about 2.5 lakh population, as the entire people were not really affected by famine. The urban areas did not feel the pinch. In fact, the effect of famine was less in Aizawl district—about 62 per cent. But in Lunglei sub-division, it was more severe. Seventy-five to 80 per cent of the landed crop was damaged which affected the entire population. But the relief procured was sufficient. In the words of an activist, 'In Aizawl town it was as if a bumper harvest was obtained. There was sufficient quantity of rice. Some of these was from as far as United States of America. It was available at the rate of Rs. 21.60 per maund and paddy at Rs. 13 per maund.'[37] The tribals of urban areas had the purchasing power. The average net income per family was Rs. 927.29 in 1960–1 whereas per capita net income was Rs. 154.55.[38] Out of a total population of 2,06,606, in 1961 only 1,25,686 or 47.23 per cent were earning while others were dependents. Out of them 35 per cent were in agricultural pursuits and hence when the harvest was damaged, they had virtually no income. The consumption expenditure was Rs. 1,029.59 showing a deficit of Rs. 1,029.59–Rs. 927.29.

Cultivators	1,09,518	Transport, Communication & Storage	742
Agricultural Labourers	33	Services	6,725
Mining Activities	535	Total Workers	1,25,686
Household Industry	5,656	Non-Workers	1,40,377
Construction	495	Total Population	2,66,063
Trade and Commerce	998		

But in the rural areas people neither had the purchasing power nor did supply reach them.

In addition to the government sanction, a sum of Rs. 1,80,000 had been released from the Assam government relief funds. Rs. 55,000 had been donated from the Chief Minister's Relief Fund. The government also collected a sum of Rs. 67,268 *p.* 41 as donations from the general public of Assam; a cash assistance of Rs. 10,000 was donated by the Indian Peoples Famine Trust Fund, New Delhi and 54,000 maunds of rice and medicines were donated by National Christian Council of India.

The relief given directly to the famine affected people by the government amounted to Rs. 8,960,000 during 1959–60 and 1960–1. Sixty Jeeps, 12 power wagons and pick-up vans were placed at the disposal of the district administration for the operation. The supply was obtained by air in the initial period followed by river ways and roadways. The famine also saw the construction of road networks, which was necessitated by relief work. It was only during the Second World War that the first jeepable road connecting Silchar with Aizawl was constructed. In 1953 Aizawl–Lunglei Road up to Zemabawk was opened. The people undertook voluntary labour for its constitution. During the famine these were the only two roads. Only clay pathways still could approach Champhai and North Vanlaiphai. The black topped roads were localized in Aizawl town only. Till 1951–2, the Aizawl–Silchar Road was barely jeepable. During and after the *mautam* the new jeepable roads were constructed which were, Silchar–Aizawl and Aizawl–Lungleh; Aizawl–Champai; Aizawl–Thenzawl; Lungleh–Demagiri; and Lungleh–Lungtlai. Except Aizawl–Silchar, others were only fair weather road. There was no river navigable, no railways and no airfields or landing strips. The road mileage increased in 1960–1 to 1486.32 km. Out of this 222.04 km were black topped; water bound metalled 8.30 km; gravelled 184.77 km; fair weather motorable roads 365.85 km; total motorable roads 559.86 km; non-motorable roads 926.55 km. Even out of 72 villages in the Mizo Hills, a sample survey of conditions of only 40 villages were taken. Out of them only 14 were connected by footpath, 4 had fair weather road and all weather roads.[39]

The administration formed a Central Famine Relief Committee with L. Ingty, Deputy Commissioner as the Chairman, J. Nampui, of the Indian Administrative Service—as Vice-Chairman. B.T Sanga, Additional Deputy Commissioner as Secretary. It had members from the Welsh Christian Mission (Reverend J.M. Lloyd), political parties (Thanglidena, Secretary EITU and Sang Zuala from the ZO Reunification Committee (ZRC) and non-governmental organizations (such as Rokhuma of the Anti-Famine Campaign Organisation. As an afterthought, the Secretary of Mizo Union was also included. It also organized village relief committees to function as its channel for famine relief. On 18 April 1960 there was a debate on the Mizo famine broadcasted though All India Radio where Prime Minister Nehru expressed his opinion that there existed a tendency to exaggerate the starvation deaths and claims were made that almost all deaths were starvation deaths. It was therefore decided that information for all starvation death should be first reported to the relief committee before it was supplied to press or higher authorities for verification.

In view of the intensive relief operations it was decided to utilize the services of state government employees stationed in Mizo Hills and accordingly these extra manpower was put in service. All the staff of the Deputy Commissioner's office were asked to donate from their salary to the Central Relief Committee. Alongside famine relief, land reclamation work was also done through tractor and bulldozers. By April 1960, many acres of land were reclaimed: Aizawl–10, Mikthun–16, Chaltlang–110, District Council Play Ground–1 in Kolasib. In Zemabawk alone 82 acres were flattened for the use of the public.

Politics of the Aftermath

As the famine subsided, politics heated up. The famine had cost a number of lives, created a lot of hardship for the people. But the greatest casualty of the famine was the Mizo Union. Since it was not given any responsibility of famine relief and the state government and the MNFF managed everything, the image of the Mizo Union was affected. Due to the anti-Mizo Union propaganda of the Assam Congress as well as the MNFF, people were alienated from the Mizo Union. At the same time due to the slow reaction of the Government of Assam, the relationship between the Chaliha government and Mizo Union-led autonomous District Council reached its lowest ebb. The Mizo Union, which was the staunchest supporter of the Congress at one time gradually drifted away and became more and more critical of the Assam government in their public utterances. The District Council felt that the Assam government did not care enough for the famine ravaged peoples of Mizoram and even bypassed the District Council on matters of crucial importance. The hostility became evident when unnecessary objections were raised like as to whether the District Council was entitled to use 'Service' stamp or whether the Deputy Commissioner's family members were entitled to occupy the inspection bungalow and other such trivial issues. Things came to a head and the Mizo Union eventually parted company with the Congress following sharp differences over the issue of famine relief and the more serious Assam State Language Act, 1960. The Mizo Union passed a resolution withdrawing its support to the Congress on the grounds that (a) the Chaliha government had not done enough to tackle the famine of 1959 and (b) the Assam Pradesh Congress Committee (APCC) had planned to declare Assamese as the state language which would give undue benefit to the plains people of Assam and would be disadvantageous to the hill people.

As a matter of fact, after the sudden drop of popularity during the famine, the Mizo Union had no option but to withdraw from its alliance with the Congress to salvage a semblance of its credibility. The issue of handling of famine relief was taken up as a counter to the propaganda that was going on against itself. The weekly confidential report of the government reflected that other parties have been consistently blaming the Mizo Union whenever there was a report of relief material not reaching a particular area, e.g. the Biate area. In 21 and 22 June 1960, the Chief Minister B.P. Chaliha announced that the Assam Official Language Bill, 1960 would be introduced in October. This sparked off fresh movement for a separate hill state in the tribal areas. Such a demand, earlier raised by Eastern India Tribal Union (EITU), had by then died down and the Act, if passed, would mean imposition of Assamese on the non-Assamese speaking population. The EITU unit of Mizoram immediately picked up the issue and the MNFF too was aggressively using the issue to communalize the situation. The Mizo Union to better its image had to do something. It was 'seriously thinking of demanding the formation of a Greater Mizoram with some portions of Cachar, Manipur state and Tripura inhabited by the Mizos'.[40] In this connection V.L.

Tluanga, C. Pahlira and C. Chuanga were to visit Manipur state to mobilize support of the Mizo people living in Manipur.[41] Meánwhile the Naga delegation was meeting the Prime Minister in New Delhi to end the vexed Naga problem. The Mizo Union was keenly observing the development. When statehood was granted to the Nagas, as a result of an agreement between the two parties, the Mizo Union was encouraged. The leaders of the party were anxious to know the implications of the state given to the Naga people.[42]

In July 1960, both UMFO and Mizo Union right wing merged with the EITU, which was now transformed into a new political party covering the entire north-eastern region, the All Party Hill Leaders Conference (APHLC). The APHLC organized a huge movement against the leadership of Assam and demanded a separate hill state for the tribals out of Assam. Thus the Congress was pitied against the might of tribal leaders under the APHLC. The language issue had created a massive crisis for the Assamese leadership. While the Bengali-speaking population throughout the state rose in a massive uprising resisting the move, the tribal leaders intensified their separatist movement threatening the integration of the state. It had already lost its only ally—the Mizo Union—during the famine. Now in all the tribal areas of the north-east there was a growing hostility against the Assamese people. In Mizo Hills too 'a feeling of dislike of the Assamese people was fast growing in the minds of local people' due to the language issue. Subsequently there were a number of cases of assault of Assamese residents.[43]

The decline of the Mizo Union in Mizo Hills following the famine created a vacuum in the district. B.P. Chaliha found it an opportune movement to counter the Mizo Union by organizing another party to fill the void. He had already observed the popularity of the MNFF. Chaliha therefore inspired its leadership to transform it into a political party. It was even suspected that Laldenga had been provided financial aid for organizing a political party out of MNFF to work against the Mizo Union which B.P. Chaliha considered 'a thorn in the flesh'.[44]

The MNFF had already earned the gratitude of the people for their work during the famine. Now it began to fulfil political ambition. It soon began to publish a daily paper *Mizo Aw* (Mizo Voice) whose editor was Laldenga himself. Although the MNFF was still a non-political body, its mouthpiece was used for political propaganda. It mainly consisted of a tirade against the Mizo Union party while at the same time it tried to rouse secessionist ideas amongst the people of Mizo Hills. Despite this, it continued to receive the patronage of the Assam Chief Minister. The MNFF received a grant of Rs. 1,500 for the publication of this newspaper from the district administration, obviously under instruction from the Chief Minister himself. On 30 August 1961 the prominent personalities of the MNFF met at the residence of Manliana, wherein it was resolved to transform the organization into a political party. They decided to re-christen it as the Mizo National Front (MNF). The word famine was dropped. The idea was to appropriate the popularity of the MNFF and thus make the new party readily acceptable. The formal announcement was made on 28 October 1961 when MNFF was declared to be a political party called MNF. Laldenga was elected its president and Vanlawma its secretary. Its formulaed objectives are mentioned as below:

1. To achieve the highest sovereignty and to unite all the Mizos to live under one political boundary.
2. To uplift the Mizo position and to develop it to the highest extent.
3. To preserve and safeguard Christianity.

Right from the beginning the party made its military and secessionist intentions clear to the Mizo people, specially the youths. It emphasized on introduction of the youth elements and even promised to give arms to each of them. The party organized their political campaign beginning with prayers. Laldenga used his great oratory to indoctrinate youth by narrating and populariszing the Mizo past in glorifying words. They were given discourses on nationalism and the preservation of Christianity from the domineering Hindu nation. A section of the youth, drivers, conductors, businessmen, ex-chiefs, ex-serviceman and anti-Mizo Union Mizo individuals were immediately overwhelmed by the objective and ideology of MNF. For them Laldenga emerged as a cult-figure. He was a devout Christian who attended the Church regularly. He was an excellent orator whose speeches were very moving, ambitious and full of promises and false hopes. In fact, during the initial period of the MNF movement, the Mizos in the rural areas considered him the 'Messenger of God'. He could thus muster a good following in both the rural and urban areas. His army background helped him in the formation of the Mizo National Army—the armed wing of the MNF. But Laldenga had a shrewd mind. Even though he made secession as an objective of the party, he did not go for an outright secessionist movement. He first wanted to earn political legitimacy for his party and gain acceptability among the masses of Mizo Hills by participating in the elections.

In the general elections of 1962, the Mizo Union contested and won two State Assembly seats while the third seat went to a Mizo Union supported nominee of the Eastern India Tribal Union, another constituent of the APHLC. Then on a directive from the APHLC both the Mizo Union nominees resigned their seats on 24 October 1962, but the EITU candidate refused to resign his seat. The by-elections for the two seats held in 1963, were contested by MNF. It proved its increasing popularity by winning both the seats contested from Aizawl West and the Lunglei constituency. In the election to the village councils held in 1963, the Mizo Union secured 228 seats against MNF's 145, the Congress had 16, EITU, 12 and Independents, 10. When the third legislature seat in the Mizo Hills fell vacant due to the resignation of Thanlhira (EITU), who was appointed a member of the Assam Public Service Commission, the by-election was won by Mizo Union President Chunga himself in 1964, although the MNF mustered all its strength to contest this seat for its candidate, P.B. Rosanga, a young commerce graduate.

Meanwhile, the MNF stepped up its activities, of campaigning, indoctrination and mobilization. It made nationalism its primary instrument of mobilization. It propagated that before the coming of the British, the Mizos were an independent nation.[46] In fact, they were a 'distinct nation created, moulded and nurtured by God and nature', and the administration of the chiefs were similar to the 'Greek city states'. They considered the MU led merger of Mizoram with India as an act of political immaturity, ignorance and absence of farsightedness.[47] The MNF leaders declared in their speeches that 'freedom' was their basic 'human right'.[48] The newspaper *Zalenna* (Freedom) under the editorship of a seasoned activist R. Vanlawma was used to circulate such ideas. Simultaneously pamphlets were also used with regularity for the same objective. MNF also brought in the factor of religion.[49] It depicted India as a land of Hindus while Mizoram consisted of Christians, which face persecution under Hindu rule. Lalthangliana alleged that the Indian officials intentionally used Sundays for their official visit of Mizo Hills, so as to render the Mizo Christians unable to observe their prayers. It asserted that the Mizos should refuse to be dominated or assimilated into the fold of idol-worshippers. He promised that under MNF rule, Mizos would be free to practise their Christian religion without any hindrances. The MNF also orchestrated occasional brawls between the Mizos and the non-Mizos to drive a wedge between the tribal and non-tribal in Mizo Hills. The MNF manifesto declared, 'Mizoram is for Mizos'.[50]

To counter the MNF influence, the MU also initiated a movement for the separation of the Mizo Hills from Assam and to constitute it into a state within the Indian union. It vowed to achieve it through Gandhian methods and recruited volunteers for this purpose.[51] The situation became explosive due to the confrontation between the MU and MNF. To avoid a possible conflict between the two organizations the neutral Mizos organized a conference at Churachandpur (Manipur) in January 1963. The MU agreed to postpone the statehood demand for Mizoram while the MNF agreed to drop its demand for secession of the Mizo Hills from the Indian union and adopt constitutional methods to achieve its ends.[52] The Churachandpur conference, however resolved to work for the integration of all the Mizo areas of north-east India into a single state. Accordingly, in October 1965, a MU delegation met the Prime Minister Lal Bahadur Shastri and submitted a memorandum demanding the formation of a Mizoram state. Shastri assured the Mizo Union leaders that he would 'have a word' with H.V. Pataskar, Chairman of the Hill Areas Commission, so that the latter examined the Mizo Union's demand while working out the administrative arrangement proposed under the Nehru Plan of Autonomy.[53] But the sudden demise of the Prime Minister and the subsequent refusal of the Pataskar Commission to consider the demand for a separate state made both MU and MNF active again. Dissatisfied, the MU boycotted the Pataskar Commission. The rejuvenated MNF submitted a memorandum to the new prime minister demanding complete freedom to the Mizos (30 October 1965).[54]

The Mizos from times immemorial lived in complete independence without interference. Chiefs of the different clans ruled over separate hills and valleys with supreme authority and their administration was very much like Greek City State's of the past. Their territory and every part thereof had never been conquered or subjugated by their neighbouring state . . . scattered as they are divided (by the British), the Mizo people are inseparably knitted together by their strong bond of tradition, custom, culture, language, social life and religion wherever they are. The Mizos stood as a separate nation even before the advent of British Government, having a nationality distinct and separate from that of India. In a nutshell they are a distinct nation, created, moulded and nurtured by God and nature. . . .

In other words, the Mizos had never been under the Indian government and never had any connection with the politics and policies of the various groups of Indian opinion. When India was on the threshold of independence the relation of the Mizos with the British government and also with British India were fully realised by the Indian National Congress leaders. . . . Due solely to their political immaturity, ignorance and lack of consciousness of their fate, representatives of the Mizo Union, the largest political organization at that time representing all political including representatives of religious dominations and social organizations that were in existence submitted their demand and chose integration with free India imposing condition *inter-alia* 'that the Lushai Hills will be allowed to opt out of Indian Union when they wish to do so subject to a minimum period of ten years. . .'.

During the fifteen years of close contact and association with India, the Mizo people had not been able to feel at home with Indians or in India nor have they been able to feel that their joys and sorrows have really been shared by India. They therefore do not feel Indian. Being created a separate nation they cannot go against nature to cross the barriers of nationality. They refused to occupy a place within India as they consider it to be unworthy of their national dignity and harmful to the interest of their prosperity. Nationalism and patriotism inspired by the political consciousness has now reached its maturity and the cry for political self-determination is the only wish and aspiration of the people *ne plus ultra*, the only final and perfect embodiment of social living for them. The only aspiration and political cry is the creation of Mizoram a free and sovereign state to govern herself to work out her own destiny to formulate her own foreign policy. . . .

Though known as head hunters and a martial race, the Mizos commit themselves to a policy of non-violence in their struggle and have no intention of employing any other means to achieve their political demand. If on the other hand the Government of India brings exploitation and suppressive measured into operation employing military might against the Mizo people as is done in the case of the Nagas, which God forbid, would be erroneous and futile for both the parties for a soul cannot be destroyed by weapons.

For this end it is in goodwill and understanding that the Mizo nation voices her rightful and legitimate claim of full self-determination through this memorandum. . . .

The situation was tense. There was dissatisfaction in the Mizo Hills with the administration of the Government of Assam. The MNF used the situation to its advantage. It attracted the younger generation on account of its radicalism and romantic idealism. Loyal Mizos foreseeing dangers advised the government to decide once for all and concede all reasonable demands forthwith in one instalment rather than allow itself to be pushed into one concession to another. Even the Governor of Assam, Vishnu Sahay, was of the same opinion.[55] But the Government of Assam was indecisive. Vishnu Sahay even suggested that the Mizo Hills should be constituted into a separate administration like Nagaland, Tripura and Manipur before insurgency erupted there. But Bimala

Prasad Chaliha the Assam Chief Minister, was totally opposed to such an idea.[56]

Meanwhile the MNF had prepared itself for a *coup de etat.* It trained its cadres and prepared for a multi-pronged attack by the members of the Mizo National Army at the same time at different government offices in different parts of Mizo Hills. The main target was the Assam Rifles concentration in Aizawl and the treasury offices of the government. The idea was to take the government as well as para-military forces completely by surprise by the armed stack and before the Government of India could send its army, the MNF would declare Independence, which hopefully would immediately be recognized by some countries. This would compel the Indian government to withdraw its forces from the Mizo Hills. Accordingly the attack was launched on 28 February 1966, which continued till 1 March 1966. About 800–1300 armed MNA cadre took part in the violent action that took place simultaneously in Aizawl, Lunglei, Vairangte, Chawngte, Chimluang, Kolashib, Champai, Saireng and Demagri. The MNF looted treasury, kidnapped government officials, killed security personnel, damaged property and set fire to the bazaar. They disrupted communication and blocked roadways to prevent the Indian Army from reaching the Mizo Hills.

Within hours of the attack, on 1 March 1966, the MNF declared independence and formed a government with Laldenga, as the president of this Mizoram *Sawrkar*, Lalnunmawia as the Vice-President, Sainghaka as the Home Minister, C. Lalkhawliana as Finance Minister, R. Zamawia as the Defence Minister, Ngurkunga the Information Minister and John F. Manliana the Chief Justice. The declaration of Independence read:[57]

In the course of history, it becomes invariably necessary for mankind to assume their social, economic and political status to which the laws of the nature's God entitle them. We hold this truth to be a self-evident that all men are created equal and that they are endowed with inalienable fundamental human rights and dignity of human persons and to secure these rights, Governments are instituted among men deriving their just powers from the consent of the governed and whenever any form of Government becomes destructive of this it is the right of the people to alter, change, modify and abolish it and institute a new Government laying its foundation on such principles and to organise its powers in such forms as to them shall seem most likely to effect their rights and dignity. The Mizos created and moulded into a nation and nurtured as such by nature's God, have been intolerably dominated by the people of India in contravention to the laws of nature.

Thus began the bloody history of Mizo insurgency in north-east India. After the declaration of Independence the MNF appealed to other nations to extend recognition to it as a sovereign nation. Although the interior areas of Mizo Hills promptly came under the control of the rebels in Aizawl, the battalion of Assam Rifles held out. The Government of Assam promptly declared the hills as a disturbed area and the army was sent to tackle the insurgency. But the same problem of communication infrastructure, which prevented famine relief to reach the interior areas hindered the access of the army to those areas. A column of 61 Mountain Brigade landed in Aizawl by helicopter. The entire town was under the control of the MNF except the Assam Rifles regimental area. As frontal combat was not possible the Indian Air Force bombed MNF positions in the town on 5 March. The bombing continued on 6 March as well. The Mountain Brigade in the meantime had reached Aizawl through Kolasib and freed from the control of the insurgents. Lunglei continued to be under the control of MNF for some more time. The civilians who had deserted Aizawl during the bombardment began to return by 7 March though sporadic skirmishes continued. The Indian forces fanned out to Lunglei and other interior areas by the fourteenth. Champai was rescued on 14 and Demagiri on 17 March. As a result of the all round attacks by the Indian Army, the MNF guerrillas had to leave the Mizo Hills and take shelter in the neighbouring country. A provisional government of the MNF started operating from East Pakistan. Its guerrillas were camped and trained in the jungles of Chittagong Hills tract from where they continued their struggle. The MNF led movement lasted exactly for a period of twenty years after which the MNF signed a peace agreement with the central government of India.[58]

Politics of Nationalism

As has been seen, the construction of the Mizo national identity was the work of a modern liberal Mizo middle class. This identity was based on cultural markers as well but was liberal, cosmopolitan and secular. It's primary objective was to attain social equality and a commensurate modernization with other free nations rather than sovereignty under a chiefly regime. But the dominant chiefs appropriated the identity to communalize and advocate a separate sovereign existence. The legitimization of a sovereign national existence was an attempt to perpetuate their autocratic hegemonic rule as they foresaw doom in becoming a part of egalitarian India. But their struggle was lost to a radical modern middle class who believed in a republican, liberal and egalitarian Indian union.

The famine of 1959 provided the chiefs an opportunity to raise their heads again. They had already lost their position and power after the union with India but continued to perpetuate their hegemony over the people. They wanted to get back to power. The famine supplied the pretext. The discourse of nationalism was the instrument. They had already constructed the otherness of their tribe. Now they used it to show that very otherness was the cause of *relative deprivation* and *internal colonialism* of the post-colonial Indian state. In general the *theory of internal colonialism* involves a two-stage argument. First it seeks to explain the causes of uneven economic development between different regional communities within a state. Second, the theory argues that such disparities are the fundamental cause of the emergence of ethnic nationalist movements amongst cultural groups located in peripheral regions. The theory developed as its name implies, as an analogy of international colonialism and the anti-colonialist nationalist responses which that evoked.[59] The MNF invoked the premises of this theory and focused on the situation of small, subordinate communities occupying peripheral underdeveloped regions within a state and on the character of the states which perpetuates deprivation and exacerbates political oppression and cultural domination of such groups. The facts of famine relief were enough to depict the relative deprivation and internal colonialism of the post-colonial Indian state. It depicted how a peripheral region and its people were thrown to the mercy of a natural disaster without any aid from government. They argued that this could happen only because the tribals were the 'other' *vis-à-vis* the rest of India in physical types, language and culture and above all religion. Through an effective direct contact programme the MNF leaders mobilized people by feeding them on their deprivation and neglect on the part of the Indian government. They reiterated that even a colonial government like that of the British was more efficacious in a similar situation in the past. It accused the provincial Assam government under whose patronage it was born, of deliberately ignoring the tribal appeal for famine relief and delaying the actual relief.

There has been lot of political discussion and academic studies accusing the Assam government of its unsympathetic response to the famine distress and thereby sparking off the subsequent insurgency in Mizo Hills. However a closer look would depict a different situation. It is true that the Government of Assam dominated by non-tribals did not take the early warnings of the Mizo Union regarding the impending famine seriously. But for this it cannot be held responsible. The Mizos were a tribe who lived in the impenetrable hills maintaining extreme exclusivity. The plainsmen never knew what was happening inside the hills. The entire knowledge of bamboo flowering and related famine in the Mizo Hills therefore remained confined to the tribe themselves. In the pre-colonial times the plainsmen of Barak and Surma valleys were aware of the distress as the tribes would migrate to the plains in search of food and shelter during the famine. But once the British took over the Mizo Hills, they not only prevented such migration to the hills, they actually stopped all normal intercourse between the tribal and the plainsmen. Once the generation of plainsmen who had some idea of the occurrences in the hills disappeared, the younger people could not provide any clue to the tribal life and environment. After the 1873 Inner Line Regulations, not only non-Mizos could not enter the hills, even Mizos were discouraged from visiting the plain areas. In fact, the historical institution of tribal-non-tribal or hill-plains interdependence was permanently shattered. Under the combined initiative of the colonial administration and Christian missionaries the tribals were developed as an exclusive people cocooned inside the hills with very little interaction with the outside world. It was a disruption of the historical institution, which blocked the movement of the hill people. The colonial administration had already paralysed the traditional economy. It did not stop to think that a people could not be permanently confined to steep hills as survival in the hard soils with no communication with the outer world would be absolutely impossible. The British had no doubt provided alternative means of livelihood and offered state aid to tackle recurring famines. This worked fine as long as the colonial administration remained in the hills. But once it withdrew, the tribals were left to fend for themselves. The state that succeeded the colonial administration had neither the knowledge nor infrastructure to mitigate the calamity. In an attempt to perpetuate the colonial mode of administration the post-colonial state too continued the policy of maintaining the 'splendid isolation' of the tribal. An inward looking tribal elite encouraged it. The result was disastrous. Neither the tribals realized how vulnerable they had become over the years of colonial rule nor did the state evolve any alternative policy. The post-colonial state of India believed in the colonial fear that any interference in tribal life would incite rebellion. Hence it played safe by continuing colonial policy. But half a century of colonial rule had not only destroyed the self-reliance of the tribal but also made them dependent on outside agencies like the state. In fact, the autonomy of the tribals and non-interference of the colonial administration in their affairs were both absolute myths. The famine had amply exposed this myth. The Mizos had lost its traditional coping mechanism with the rat-famine and were dependent on external agencies for their survival during calamities. The post-colonial state had granted them autonomy in the form of the District Council under the Sixth Schedule of the Constitution but without any financial powers it was meaningless. For a mere Rs. 15,00,000 it had to depend on the provincial state. The state administration took advantage of this dependence by initially rejecting the demands of the District Council only to demonstrate its power to one of its adversaries—the Mizo Union party. But when it actually realized the seriousness of the problem it went all out to mitigate the crisis. But once again it realized the problems the policy of 'iron-curtaining the tribal' had resulted in. With all the resources at its disposal, the administration could not ease the crisis as it had neither the infrastructure nor the knowledge to reach the affected people. It granted Rs. 190 lakh on famine relief though the affected people wanted only Rs. 15,00,000. Yet the state was accused of indifference. It was because the post-colonial government had not anticipated the extent of reliance of the Mizos on administrative apparatus. The famine showed the extent to which the Mizo people were dependent in external agencies. The MNF took advantage of the situation. Interestingly it distributed state relief materials to the affected to people to take credit of being pro-people and at the same time discredited the same government as anti-tribal for delaying relief. It contrasted the post-colonial Indian government with that of the colonial and pre-colonial chiefly order for their role in the mitigation of previous famines and campaigned for sovereignty on these planks. In other words it favoured independence but actually talked of reverting to those orders where people were dependent on the chiefs or colonial administrative machinery.

The hostility towards the Indian state and polarization between the tribal and the non-tribal were completed through a successful campaign, which often turned violent. The MNF was successful in projecting the Indian state as a *vai* (alien) and its people anti-tribal. As a consequence of this propaganda even the ordinary non-Mizo settlers were made an object of hostility. They were targeted, assaulted and even asked to leave Mizo Hills.

Then the nationalists endeavoured to turn this otherness of the Mizos *vis-à-vis* the Indian into a comprehensive self, the Mizo national self—the *Hnam*. It projected the MNF as the agent of Mizo liberation from the colonial subjugation of India and the goal was the establishment of a Mizo nation state—Mizoram. The prospective freedom was not just from occupationist India but 'the freedom to pursue happiness, freedom from feeling of insecurity and freedom from ignorance, poverty and wants'. The nationalist project stood for the unification of all the cousin tribes to eventually establish a Greater Mizoram or Zo-land. It would thus neutralize the policy of the Indian state to divide and weaken the Mizo people. It also showed that India was a polytheist Hindu state where even Muslims were incompatible. Hence a tiny Mizo people who were minority Christians were likely to be persecuted. The Mizo nation state would be the ideal polity for the Mizos where the language would be Mizo and Christianity the state religion. No non-Mizos would live in that territory. The new nation state would 'improve the social, economic and political condition of the Mizos'.

The post-famine politics in Mizoram witnessed strange development. The Mizo Union, which reflected peoples aspirations in the period of Indian Independence and accordingly secured the merger of the Mizo Hills with the Indian union suddenly lost out. They lost out even though they were the first to raise the impending danger of a periodic famine, the first to officially seek advance financial assistance to combat famine distress, the first to oppose linguistic expansionism of the provincial Assam government as well as the imposition of Hindi by the Government of India and the first to demand a separate autonomous unit (state) for the Mizo people. It lost out to a relatively new political party, MNF, which represented the old adversaries of the Mizo Union. MNF had then lost the battle but now won the war basically by appropriating the nationalist discourse, which the Mizo Union failed to do. The union recognized Mizo ethnicity but could not demand secession. The MNF on the other hand mobilized people on the slogan of '*I told you so*'. It had already warned the people against merging with India as the two represented opposite national identities. It forewarned that the big nation would soon turn oppressive and under whom the rights and identity of a small Mizo state was not secure. They now returned to the same slogan stating that the neglect by the Indian state confirmed their old suspicion. The two nations were incompatible and should therefore part with each other. From construction of nationhood to presentation of India as oppressor nation to the demand of the right to self-determination, the MNF used all features of nationalist discourse on a famine-ridden people rendered vulnerable by its ravages. The picture of an affluent, famine-free, plentiful, self-ruled nation raised their hopes. For an exhausted population, the prospect was rosy indeed and a fertile ground for the cultivation of nationalist ideas. What the dominant chiefly class could not achieve in the 1940s was realized in the 1960s using the same rhetoric. A natural disaster made all the difference.

NOTES

1. For details of this see Sajal Nag, *India and North East India: Mind, Politics and the Process of Integration*, Delhi: Regency, 1999.
2. Resolution of the Working Committee of Indian National Congress on United India and Self-Determination, 12–18 and 21–4 September 1948.
3. R. Vanlawma, *Ka Ram Leh Kei* (in Mizo), Aizawl: Author, 1972, p. 119, cited in R.Sena. Samuelson, *Love Mizoram*, Imphal: Author, 1985, p. 30.
4. Ibid.
5. Vanthuama cited in Samuelson, op. cit.
6. Ibid.
7. Pachhunga, Dahrawka and Hmartawnpunga, *Independence*, Aizawl, 29 July 1947.
8. D. Ronghaka, *Zoram Independent*, 5 May 1947.
9. It was dated 18 June 1947.
10. Nag, op. cit., p. 90.
11. K. Zawla, *Zoram Din Hmun Dik Hmuh Chuah Thelhna Tur*, Aizawl, 7 June 1947.
12. Samuelson, op. cit., p. 47.
13. B.B. Goswami, *Mizo Unrest: A Study of Politicisation of Culture*, Jaipur: Alekh, 1979, p. 136.
14. Ibid.
15. Ibid.
16. Mizo Zalenna Pawl: Pawl Din Dan (Constitution of United Mizo Freedom Organisation), Aizawl, 1947.

17. R.N. Prasad, *Government and Politics in Mizoram 1947–86*, Delhi: Concept, 1987, p. 80.
18. Vanlawma, op. cit., p. 217.
19. Goswami, op. cit., p. 136.
20. R.S. Samuelson, op. cit., pp. 47–50.
21. Ibid.
22. K. Laldinpuii, 'The Mizo Non-Cooperation Movement', *Proceedings of the North Eastern India History Association*, Aizawl Session, 1996, pp. 353–66.
23. Ibid.
24. Nari, Rustomji, *Enchanted Frontiers*, Delhi: Oxford University Press, 1973, pp. 96–7.
25. During the anti-slavery movement led by Dr. Fraser, the administration preferred to throw out the missionary rather then anger the chiefs.
26. Report of the Enquiry into the Chiefs Position, by B.M. Roy, Asst. Suptd. Lushai Hills, dated 16 May 1940, MSS EUR/ E 361/27, BL.
27. C. Rokhuma, *What is the Anti-Famine Campaign Organisation Doing?* Aizawl: Author, 1988, p. 121.
28. Ibid., p. 125.
29. Ibid., p. 127.
30. S.K. Chaube, *Hill Politics in North East India*, Delhi: Orient Longman, 1973, rpt. 1999, p. 179.
31. V. Venkata Rao, *Century of Government and Politics in North East India*, vol. III, *Mizoram*, Delhi: S. Chand & Co., 1999, p. 236.
32. Chaube, op. cit.
33. S.H. Pautu, 'Separatist Politics in Mizo Hills', unpublished M.Phil. thesis, North-Eastern Hill University, Shillong, 1982. Chaliha favoured MNF not only during the famine but also later when the MNF chief Laldenga was caught for anti-national activities. When Laldenga was arrested for travelling to Pakistan in a conspiracy to secure the secession of Mizo Hills, Chaliha released him. For this he faced severe attack in the Assam Legislative Assembly. See *Assam Legislative Assembly Proceedings*, 5 March 1966, pp. 466–91.
34. Chaube, op. cit.
35. Goswami, op. cit., pp. 143–4.
36. Ibid.
37. Rokhuma, op. cit.
38. *An Interim Report on the Socio-Economic Survey of Mizo District 1962*, Office of the District Statistical Officer, Aizawl, 1963.
39. Ibid.

40. *Weekly Confidential Report for the Week Ending 6 August 1960*, MSA.
41. Ibid.
42. Ibid., *21 August 1960*, MSA.
43. Ibid., *13 August 1960*, MSA.
44. C.Z. Verghese and R.L. Thanzuwma, *A History of the Mizos*, vol. II, Delhi: Vikas, 1997, p. 13.
45. Goswami, op. cit., p. 143.
46. *Memorandum to the Prime Minister of India by MNF*, Aizawl, 30 October 1965.
47. Ibid.
48. Ibid.
49. Goswami, op. cit., pp. 149–50.
50. J.V. Hluna, *Church and Political Upheaval in Mizoram*, Aizawl: Author, 1985, p. 92.
51. Rao, op. cit., p. 238.
52. Ibid.
53. Ibid.
54. As in note 42.
55. Rao, op. cit., p. 240.
56. Ibid.
57. *MNF Declaration of Independence,* 1 March 1966, Aizawl.
58. For details of the insurgency see Sajal Nag, *Contesting Marginality: Ethnicity, Insurgency and Subnationalism in North-East India*, Delhi: Manohar, 2002.
59. N.B. Chaloult and Y Chaloult, 'The Internal Colonialism Concept: Methodological Consideration', *Social and Economic Studies*, 1979, vol. 28, no. 4, p. 87. J.L. Dove, 'Modeling Internal Colonialism: History and Prospect', *World Development*, 1989, vol. 17, no. 6. Robert J. Hind, 'The Internal Colonial Concept', in *Comparative Studies and Society and History*, vol. 26, no. 3, 1984. A. Orridge, 'Uneven Development and Nationalism: 1/2', *Political Studies*, vol. 29, no. 2. 1981, p. 190. Michael Hechter, 'Internal Colonialism Revisited', in E.A. Tirykian and R. Rogowski, eds., *New Nationalisms of the Developed West*, Boston: Allen and Unwin, 1985. Harold Wolpe, 'The Theory of Internal Colonialism: The South African Case', in Ivan Oxaal and D. Booth, eds., *Beyond the Sociology of Development: Economy and Society in Latin America and Africa,* London: Routledge, 1975. M. Mechter and M. Levi, 'The Comparative Analysis of Ethno-Regional Movements', *Ethnic and Racial Studies*, vol. 2, no. 3, 1979.

CHAPTER 6

Conclusion

'A famine is shortage of food so extreme and protracted as to result in widespread persisting hunger, notable emaciation in many of the affected population and a considerable elevation of the community death rate attributable to starvation' said the *International Encyclopaedia of Social Science*. Famines are not new to history but its character and causes have changed over time. In pre-modern times famine affected the entire population of an area but in modern times it affects farmers and poor. Shortage increases the price of food and people with ample purchasing power buy them and even stock them for future causing further scarcity. Thus families with low income feel the pinch of hunger and starvation and are even forced to migrate to food abundant areas.

Famine has many causes. In the nineteenth century, Walford listed twelve causes classifying them into natural causes beyond human control and artificial causes within human control.[1]

Of recent times however men have learnt to cope with causes of natural calamities and thereby minimize the impact of famines. Natural causes include draught, excessive rains and flood, unreasonably cold weather, depredation by vermin and insects like locust and plant diseases. Colonial States around the world have often manufactured calamities like famine artificially. The artificial causes include warfare, wartime strains and revolutions. Great famines of the world have been due to natural causes but intensified due to political factors. Whether natural or artificial, famine is always local or regional, never worldwide or continent wide or even nationwide. The major famines in history were the Irish famine of 1845-52, in Greece and Holland during the Second World War, in Persia, 1871, in Asia Minor (1874–5), Egypt (1897),

Brazil (1877), Morocco (1977–88). In this period there were no fewer than 10 famines in Persia, in India 13, including the Great Bengal Famine of 1943. China had a number of famines. The severe ones were in 1877, 1878, 1919–20, 1929–30. In fact, Russia, India and China over the past centuries have encompassed the outstanding famine areas of the world. In India famines of the last two centuries have inevitably revealed colonial reasons rather than natural causes.

North-east India has been known as the region of calamities. The devastating floods, greatest of earthquakes and epidemics like malaria and kala-azar have been an integral part of its history. North-east India has also been a zone frequented by the ravages of famine. The state of Manipur has seen devastating famines which sparked off social movements. Manipur has not been a food surplus, but food sufficient economy. But during the Second World War, unscrupulous non-Manipuri business and merchants had brought up the entire harvest for supply to the war time needs of the army elsewhere. This resulted in food shortage in the tiny state. Famine and starvation forced the Meitei women organize a movement famous as the *Nupilan* movement. The Mizo Hills have been frequented by famines caused by strange natural causes. This famine was periodical—every 30 and 50 years, which occurred due to the flowering and fruition of bamboo plants, which invariably increases the rodent population. It was caused by the depredation of food crop by these rodents. Since famine in Mizoram was periodical, it could be predicted. But despite prediction, the suffering and hardship of the people were not lessened until recent times. Famines like other calamities are therefore part of the history of Manipur and Mizoram. As such famines have often been an environmental event around which cultural, social factors were woven. In colonial times another dimension was added to this environmental event. It was the pivot around which a lot of political activity was organized. In this project we concentrated on the politics of philanthropy in Mizoram during colonial to the post-colonial period.

As the British were using their colonial might to conquer the Mizo tribal who frequently disrupted the tranquillity of the adjacent British territory by raids, plunder and kidnapping and threatening the closure of their booming tea plantation industry of the region, despite their modern military might the British were finding it difficult to match the guerrilla tactics of the tribal fighting in hard mountainous terrain. It was only once that they found the tribal submitting on their own—during a famine when hunger forced the tribal to beg food from their staunchest enemy—the British. The British on enquiry came across an interesting discovery that the submission of the tribes was due to a famine that was caused by the depredation of the entire standing crop in the field as well as the stored grains by the rats. A much interesting discovery was that this sudden invasion of the crop by the rats was because of an enormous increase in their population, which was in turn due to a periodical flowering and fruition of certain bamboo species, which cover the Mizo Hills. The consumption of bamboo fruits made the rats astonishingly prolific breeders causing rise in population. The available food in the jungle was not enough to sustain this huge population and they attacked the human population when they exhausted the jungle foods.

It was during such a famine which came periodically in the Mizo Hills that the British found that their fierce adversaries were laying down arms and settling for peace against food. The British had initially no intention in conquering these primitive tribals living in hills which offered the British no economic prospect in terms of minerals, commercial transactions, resource generation or even tea plantations. Their initial plan was to punish them for their raids and plunders in such a way that they could never disturb British territory and remained confined to the hills. But about fifty years of gruelling invasion to the hills convinced them that the only way to keep the Mizos confined to the hills was to conquer them fully and bind them by an administrative set up. The famine of 1881–2 provided an opportunity to do so when the tribals were either willing to give up arms even though temporarily or seeking the favour of their adversaries. The British decided not to conquer them militarily during the famine when the Mizos were the weakest but earn their goodwill and gratitude by showing humanitarian concerns, help them during their time of crisis and provide food to save from starvation deaths. By doing so they could easily conquer them emotionally. This would also depict that the British were no enemy as was the perception but a friend in need with paternalistic concerns. This could win over at least a section of the Mizo people. This would weaken their resistance to the British and even might ensure the moral capitulation to the British.

Secondly, even if did not achieve this objective, this provided the British with an opportunity to penetrate deep into the Mizo Hills, acquire knowledge about the terrain, the character of the people and their vulnerabilities. Such knowledge would help them decide about the future course of action either military or political. The famine relief operation actually exposed the so-far-closed and mysterious hills to the British. They were provided unrestricted access deep into the hills. The British could thus set-up markets, trade marts, storage godowns and even police outposts in strategic points. They encouraged the Mizos to become independent on these centres of exchanges. Thus British could not only post its own men in strategic locations but also ensure that the tribals became dependent on the trade marts. It meant even if the hostility did not cease, the tribal could be captured here. This also exposed the British as well as the plainsmen to the Mizos and increased their interaction through which there was possibility of the improvement of relationship. This would change the perception of each other and reduce tension.

Thirdly, once the trade marts and roads network was established, the British could think of setting an administrative set-up in the conquered area from where it could control the conquered tribals and then conquer the rest of the tribes. The British then became ambitious and thought of extending the British Empire up to the last frontier.

Fourthly, the famine introduced the British to another new idea—food for work. Since the Mizos were not averse to render labour service to the British even against payment there was dearth of a ready labour force to do the construction work for British. It was required to construct office buildings, residential bungalows and quarters for the British and their functionaries, road construction, water supply system and so on. During the famine the British asked the Mizos to render their labour against the relief provided which the starving Mizos could not pay for. In other words, food and famine relief was not free. The Mizo had to render labour services against it. By these services the Raj was able to develop Aizawl and Lungleh villages as modern townships with necessary amenities. It did not cost them anything. At the same time they earned the gratitude of the people as benevolent masters.

Lastly, the successive famines had broken the traditional self-coping mechanism of the tribal against such calamities and made them totally dependent on the British administration, a dependence that they could not break ever after. A self-sufficient people was thus transformed into permanent dependents.

The white missionaries who came with a civilizing mission in Mizo Hills found it immensely difficult to introduce Christianity among the Mizo people in the initial period. The missionaries were ridiculed, resisted and even threatened. But the famine provided an opportunity to them to influence the people to convert. They started a massive relief showing a humanitarian activism. But this humanitarianism had a politics. Like the political conquerors, the missionaries too wanted to have the gratitude and goodwill of the people through humanitarian work. They arranged relief, assisted in the relief distribution, provided medical facilities to the sick, provided employment to the able. Like the administration, they also got their churches and residences and gardens constructed against food for work. Once people were convinced of the good intentions of the missionaries, it was easy to influence them ideologically. The missionaries propagated that the practise of animism brings famine. Moreover it was visible that the people affiliated to the church or Christians had the church organization behind them in times of distress. Each famine was followed by an increase in the number of Christians. The humanitarian politics of the missionaries was successful in their objective of proselytization. In doing so they institutionalized the dependence of the Mizos on the colonial state. Thus while it served their own objective they also inadvertently allowed themselves to be the tools of colonialism.

In the post-colonial period, famine was used to practice the politics of nationalism. In pre-Independence period a group of traditional leadership failed against a modern leadership to secure the Independence of Mizoram and perpetuate their rule. The Mizo Union represented the modern leadership and while the traditional leadership (chiefs) were in two organizations, Mizo Union Right Wing and UMFO. The Mizo Union articulated the peoples' aspirations and demanded the abolition of the oppressive institution of chieftainship. Since it was possible only in a republican India the union opted to merge with India. But the traditional leadership was looking for an opportunity to strike. It came during the famine of 1959. Mizo Union by virtue of its popularity were controlling the District Council. But its image suffered due to alleged nepotism in land allotment. Despite being an ally of ruling Congress party in Assam it hosted the conference of EITU, which demanded the disintegration of Assam and the formation of a separate tribal state. This enraged the Congress Chief Minister B.P. Chaliha. He tried to develop a counter to the popular Mizo Union by patronizing another organization (MNFF). The Assam government ignored the appeal of the District Council to sanction help for famine relief and even when it provided relief it used MNFF as the distributing agency rather than the Mizo Union. MNFF seized the opportunity to harp on the neglect by Assam as well as the Indian government towards the Mizo people in times of crisis. It tried to generate Mizo nationalism as against Indian nationhood. It also blamed the Mizo Union as pro-India and bagged the entire credit of famine relief. Against the rise of an aggressive Mizo Nationalism the Mizo Union found itself helplessly sidelined. Although the Mizo Union protested against the attempt to impose Assamese language on the tribals, the MNF now transformed into a political party again through the patronage of Chaliha, depicted it as a confirmation of its fear of Hinduization. When the Mizo Union supported the idea of tribal state, MNF went a step further and demanded Mizo sovereign state. MNF was also responsible for the communalization of Mizos and non-Mizos in the hills. The MNF did not hide its agenda. In its party manifesto it openly brought out its pro-Independence agenda—the establishment of a sovereign Mizoram. It constructed a separate national identity for the Mizos, and demanded the right to self-determination. It also mobilized the alienated sections of the people behind the project. Although initially it participated in the democratic process of India, it secretly pursued its hidden agenda. Once its preparation were over it sought to implement its project. It organized a coup and tried to capture power. Thus began a long movement of insurgency in the hills for the realization of nationalist dream.

To counter the Mizo nationalist politics and dilute the alienation, the Indian state stepped in. It granted the Mizo's a union territory and when the next periodical famine struck, the state took utmost care to relieve the distress of the people. A *thingtam* (famine) hit the hills in 1977. Since the conversion of the Mizo Hills into a union territory in 1972, a deliberate policy of discouraging *jhum* cultivation was followed, as the practise was detrimental for environment. As *jhum* land shrank, the agricultural yields also dwindled. 1977 happened to be one such low harvest season, which is why the pinch of the famine was severely felt. But the central government took no chance. The government set-up a Rodent Control Committee and sent an expert team to study the situation. The government took every possible step to ensure availability of adequate foodstuff. The non-governmental organization cooperated with the administrative machinery in distributing the relief material to the needy. The 1977 *thingtam* weaned only with a single death.

As per their predictions, after the last famine in 1977, they were anticipating one more in 2007. True to their anticipation, the tall bamboo plants had already started to bloom with mauve, yellow and crimson coloured flowers all over the southern Mizo Hills as early as 2002. This time, it seemed, the calamity was likely to spread to newer areas like Arunachal Pradesh and Manipur. By 2003 hundreds of acres of paddy fields in at least 50 villages of upper Subansiri district of Arunachal Pradesh bordering China had been devastated by teeming rats in just about a fortnight. The villagers watched this devastation helplessly as they had never seen so many rats at one time. They were now seeking divine intervention from *Donyi–Polo,* the Sun–Moon god of the indigenous faith and were holding regular prayers to prevent the imminent famine.[2]

Similarly the villagers in the Tamenglong districts of Manipur were already facing acute food shortage due to the destruction of the standing crops by the invading rodent population. A total of 62 villages have been affected by the menace. The agriculture department of the state had provided 485 kg of Bromodialon (rat poison) till August 2003 without much impact on the rodents. The attack of the rats was expected to continue for the next few years as the Centre had already declared the period 2002–7 as the period of bamboo flowering and if the present situation continued the district would face a famine situation very soon.[3]

The indications were ominous and the Mizoram government, traumatized and wise by past experiences had already pressed the emergency button. Interestingly this time it was the MNF—which was born out of 1959 famine—that was in power in Mizoram. After the 20-year long insurgeney the MNF settled for a negotiated place. It then transformed itself into a political party again and came to power through a democratic process. The MNF administration which had earlier appropriated the famine now started preparations on a war footing to combat the calamity. It has set-up a high level State Rodent Control Committee with Chief Minister Zoramthanga heading it and with C. Rokhuma as his deputy in the committee.

The committee, as a first step, started a campaign among the people for mass killing of rats in the paddy fields and houses to pre-empt their proliferation. As referred early the government devised a two-pronged strategy to fight the impending menace. One, the control of the rat population and two, to store adequate food grains to meet any contingency resulting from the destruction of crops by the rats. This was to be complemented by a massive supply network developed for uninterrupted distribution of food grains to the needy.

The Indian state also did not take any chance. It was determined to lead the famine mitigation initiative. The experts from the Government of India recommended that during the bamboo flowering period the Mizo farmers should cultivate items which the rats do not eat, e.g. maize which will prevent a famine. Another expert recommended that there should be early harvesting of bamboo plants before the plants start to rot. This would prevent the crisis.

The interest shown by the central government in the imminent rat-famine demonstrated the insecurity of the Indian state *vis-à-vis* its constituents in the north-east. The region is already plagued by secessionism, insurgency, autonomy movement, and ethnic conflicts, each of which challenged the Indian state. It earnestly wanted to prevent another challenge from Mizoram, which since the Mizo Accord (1986) has remained peaceful. It showed that the Indian state has learned lessons from the past experience and wants to prevent a recurrence of the Mizo challenge which sparked off from the 1959 famine. It went to enormous extent to prevent the famine fearing that a 1959 like situation does not arise. But 1959 and 2007 are two different eras, nor are Mizos are the same. The state of Mizoram is a special category state which receives 90 per cent of its finances from the Centre. The people of Mizo Hills are hardly dependent on agricultural output. It does not produce even 50 per cent of its food requirement and is totally dependent on the supply by the Government of India. This can be seen from Tables 6.1–6.4.

It can be seen from the above tables that neither Mizoram produces its entire food requirement nor any great part of Mizo population is entirely dependent on agriculture for their sustenance. Therefore if 'famine' is the academic version of food shortage, the state of Mizoram is in a perpetual state of famine. One has only to examine the extent of its dependence on the centre for food supply to understand it. In fact, food shortage and dependence on outside agencies like the state apparatus is a colonial legacy in Mizoram. Like any tribal economy the Mizos too produced bare minimum. The slash and burn cultivation in the rocky slopes of the mountains did not permit any surplus

TABLE 6.1: THE AREA UNDER RICE PRODUCTION*

Year	Total area (ha)		Percentage share of *jhums*	
	Jhums	Valleys	Rice area	Rice output
1980–1	55,264	6,923	89	63
1981–2	55,858	7,220	89	74
1982–3	52,484	7,195	88	81
1983–4	41,281	6,390	87	71
1984–5	43,378	7,486	84	75
1985–6	45,920	9,092	83	85
1986–7	38,893	9,452	80	72
1987–8	37,803	10,661	78	72
1988–9	36,616	12,772	74	62
1989–90	38,349	14,640	72	59
1990–1	36,716	14,607	72	55
1991–2	39,175	16,464	70	54
1992–3	43,858	17,439	72	58

Source: Directorate of Agriculture, Govt. of Mizoram, Aizawl.

Note: *Cited from Daman Singh, *The Last Frontier: People and Forests in Mizoram*, TERI, Delhi, 1996.

production. The deficit was substantiated by hunting, trade with the plains and even plunders. But the colonial state had succeeded in paralysing even that subsistence producing by discouraging slash and burn agriculture, hunting, raid and plunder as it considered these as features of 'savagery'. After the initial period direct trade with the plainsmen was totally halted as the administration did not want tribesmen to come down to the plains. Professional traders from the plains were given the monopoly to trade. A number of the Mizos were employed in administrative and ecclesiastical structures which became a symbol of status and social mobility. There was soon a tendency to give up agricultural pursuits, search for white collar employment. The manpower short tribal economy was sure to feel the pinch. The cumulative result was disastrous. The Mizo tribal were totally dependent on the colonial apparatus for its food security as well as crisis mitigation which the colonial state did effectively. What is intriguing that

TABLE 6.2: TOTAL RICE PRODUCTION* AND PER CAPITA AVAILABILITY

Year	Area (ha)	Output (mt)	Per capita output (kg)	Deficit (Percentage)
1974–5	47,500	28,490	76	58
1975–6	49,201	31,164	80	56
1976–7	73,000	28,462	70	62
1977–8	68,000	16,310	39	79
1978–9	64,100	21,000	48	74
1979–80	63,200	20,440	45	76
1980–1	62,187	59,633	126	31
1981–2	63,078	30,000	61	67
1982–3	59,679	45,953	90	51
1983–4	47,671	36,713	70	62
1984–5	59,864	40,894	75	59
1985–6	55,012	64,780	115	37
1986–7	48,344	86,745	118	36
1987–8	48,464	49,227	82	55
1988–9	49,388	53,000	85	54
1989–90	52,989	59,236	92	50
1990–1	51,323	63,794	96	48
1991–2	55,639	70,974	103	44
1992–3	61,297	83,954	118	36

Source: Area and Production from Directorate of Agriculture, Govt. of Mizoram, Aizawl.

Notes: Population extrapolated from census data, per capita requirement assumed as 183 kg/year.

* Cited from Singh, op. cit.

such food insecurity has transcended the colonial times and remained a reality even in post-colonial times. The problem lay in the fact that both the post-Independence Mizoram state government as well as the Centre has failed to convincingly allay the fear of the Mizos that such an eventuality is unlikely and even if the predicted famine of 2007 were to become a reality, the Government of India is perfectly capable of feeding the tiny 6 lakh Mizo population from its reserves. The hilly state of Mizoram is now dotted with Food Corporation of India stores right into the

TABLE 6.3: GROSS CROPPED AREA, 1911–91*

Year	GCA (ha)	Percentage to total area
1911	38,228	1.81
1921	35,352	1.68
1931	31,652	1.50
1941	30,912	1.47
1951	57,893	2.75
1961	50,518	2.40
1971	84,626	4.01
1980	86,125	4.09
1991	97,540	4.63

Source: 1911–61: *District Census Handbook*, 1961; 1971: *Mizoram District Gazetteers*, TRI, 1979, 1980, 1991: Directorate of Agriculture, Govt. of Mizoram; 1981 data not available.

Note: * Cited from Singh, op. cit.

interiors. There is a well developed network of road and communication system along with air connectivity. Hence supply of food grains to the affected people would not pose much of a problem. The fear of a repetition of a famine-connected uprising in 2007 also betrays the lack of time and space in the Indian think tank. Instead of trying to attempt an all round development of the people, it is concentrating on preventing a famine which is anyway unlikely. This new 'concern' of the Indian state for the Mizo people is therefore another 'politics' to check prospective fissiparous tendency round a famine that might never come. More shocking is the policy of doling out monetary sanctions to the state government for tackling a possible famine. It has encouraged intense competition to share the booty and bred corruption. In a recent academic gathering held in capital Aizawl (19–20 April 2007), a University Professor requested a BBC correspondence, present in the Seminar, to kindly investigate and expose the embezzlement of funds sanctioned for famine mitigation in public interest. Indeed there are widespread dicontent and allegation about the misutilization of funds and its diversion to the cadres of the ruling party. On 19 April 2007 there was a violent demonstration by the farmers of the state protesting against the alleged government squandering of money sanctioned by the Centre (channelled through BAFFACOS, a state finicial agency)

TABLE 6.4: SELECT FEATURES OF URBAN CENTRES IN MIZORAM*

Urban Centre	Population (1991)	Percengage to main workers		Background
		Govt.	Business	
Aizawl	1,55,240	48	36	Ex-Dist. HQ market, Capital
Lunglei	34,609	44	15	Dist. HQ, market
Champhai	20,809	19	12	SD. HQ, WRC, ex-GC
Serchhip	13,688	20	10	Block HQ, ex-GC
Saiha	13,669	44	17	Dist. HQ
Kolasib	13,482	27	17	SD. HQ, ex-GC
Saitual	8,402	13	10	Ex-GC
Khawzawl	7,104	15	7	Block HQ, ex-GC
Vairengte	5,607	23	13	Market, ex-GC
Hnahthial	5,548	25	9	Block HQ, ex-GC
N. Kawnpui	5,290	13	9	Ex-GC
Thenzawl	4,502	12	12	Ex-GC
Darlawn	3,609	19	9	Block HQ, ex-GC
Mamit	3,546	17	8	SD. HQ, ex-GC
Sairang	3,527	19	17	Market, ex-GC
Zawlnuam	3,455	22	11	Block HQ, ex-GC
Tlabung	3,409	41	15	SD, HQ, ex-GC
N. Vanlaiphai	2,804	23	6	WRC, ex-GC
Bairabi	2,421	12	6	Rail head, ex-GC
Biate	2,325	16	3	
Khawhai	2,102	7	2	Ex-GC
Lengpui	1,808	11	6	Model Jhumia resettlement colony, ex-GC

Source: Primary Census Abstract, 1991.

Notes: Dist. – District, HQ - Headquarter, SD - Sub-division, WRC - Wet rice cultivation, GC - Group centre

*Cited from Singh, op. cit.

TABLE 6.5: DISTRIBUTION OF TOTAL MAIN WORKERS BY OCCUPATION*

Category	Persons	Male	Female
Cultivators	1,78,101	94,877	83,223
Agricultural labourers	9,527	6,181	3,346
Livestock forestry, fishing, hunting and allied activities	3,317	1,904	1,413
Mining and quarrying	631	604	27
Manufacturing, processing, serving and repairs in houehold industry	2,958	1,997	961
Manufacturing, processing, servicing, and repairs in other than houehold industry	4,606	3,844	762
Construction	7,158	6,923	235
Trade and commerce	15,078	6,285	8,793
Transport, storage, and communication	3,304	3,257	47
Other services	65,637	52,138	13,499
Total	2,90,317	1,78,011	1,12,306

Source: *Census of India 1991, series 17, Final Population Total, Mizoram.*
Note: * Cited from Singh, op. cit.

to combat the catastrophe of bamboo flowering and impending famine which turned violent. As a result 29 farmers and a policeman was injured in the ensuing encounter. It was unprecedented in the history of Mizoram. Mizoram neither had any distinctive class called farmer as everyone was one, nor had they ever politically activated themselves. In a historical move the farmers across the state formed an umbrella organization called *Zoram Kuthnathawktu Pawl* (Mizoram Farmers Association) and led a protest against the squandering of money sanctioned by the Centre for famine affected farmers. They were severely critical of the Government's policy of introducing turmeric and ginger in place of paddy as the latter crop was bound to fail during the bamboo flowering years. It warned the Government to immediately 'stop its corrupt practices or else face violent consequences'.[4] The *Zoram Kuthnathawktu Pawl* has been the product of the politics around the prospective famine of 2007. The bamboo flowering and the consequent rat invasion had started as early as 2002. Since then the agriculturists were severely affected in Mizoram. But the Government was busy in formulating ridiculous polices like mass killing of rats and premature felling of bamboo vegetation. The government even proposed that the farmers produce turmeric and ginger (which are not eaten by rats) in place of paddy. The government promised to provide financial assistance to this endeavour and buy up the products at higher rates. But the turneric seeds turned out to be of poor quality and that too was distributed selectively, allegedly only to the ruling party supporters. The turmeric harvest was poor and the ginger harvest was refused to be bought back by the government. Hence instead of the promised Rs. 10 per kg, ginger had been disposed off to private traders for Rs. 3 to 5 per kg. The farmers were suffering from bamboo flowering related crisis from 2002 to 2007 while the state government was proposing schemes after schemes as mentioned in the introductory chapter. Eventually the poverty-stricken desperate farms formed their umbrella organization at Saitual village in 2006 and within a year more than 300 branches of it were formed across the state to protest the government's failure against famine mitigation. The alienation was not confined just to the farmers alone. During the author's field study trips in Mizoram he sensed widespread discontent. Almost all sections of the population were unhappy with the way a prospective famine was being politicized and its funds were being misused. It was alleged that the ruling party was utilizing the central funds in favouring and strengthening its cadres. It was also displeased that the central government is only sanctioning funds without constituting any mechanism to monitor its utilization. Indeed, there is a simmering opposition over the issue of squandering funds meant for famine relief across all sections of the Mizo population. If these are any indications, the much publicized famine mitigation programmes of the present Mizo National Front government might actually be its nemesis. One famine (1957) created the Mizo National Front, will the other (2007) cause its exit remains to be seen.

NOTES

1. Cornelius Walford, 'Famine of the World: Past and Present', *Journal of Royal Statistical Society*, vols. 41–2, 1878–9, pp. 79–265, 433–526.
2. *The Shillong Times*, 1 September 2003, p. 3.
3. Ibid., 12 September 2003, p. 1.
4. 'Farmers' Unrest in Mizoram', in *North East Sun*, 15 May 2007, p. 10.

Glossary

Buman:	a variety of rice
Bia:	a variety of sticky rice
Bawi:	slave
Butang:	a variety of paddy
Chhippa:	Mizo equivalent of the month of June
Chakalai:	name of a ritual to drive away the famine
Chanchins:	newspaper
Dawr:	foothill opening
Donyi:	Polo—the Sun-Moon God of certain tribes of Arunachal Pradesh
Dulien:	a dialect of the Mizos
Huai:	spirit or ghosts
Hakai:	a root resembling yam
Harhna:	revival
Inpuisung:	a category of slave
Jui (*Assamese*):	fire
Jhuming:	slash and burn cultivation
Karbari:	trader
Kathis:	a measure for rice
Kuthnathawk:	farmer
Kachu:	yam
Kraws sipai:	soldiers of the Cross
Lalpa:	chief of chiefs or god almighty
Mautam:	famine resulted by the flowering ofmau species of bamboo
Muli:	a species of bamboo
Magh:	a variety of rice
Mukh:	mouth or opening
Muharrir:	accountant
Minpui Kum:	great landslides
Nupilan:	womens' movement
Pathian:	Mizo supreme god
Pani:	water
Phulrua:	a variety of bamboo
Pern:	migration

Puithiam:	Mizo village sorcerer
Punji:	village
Puma Zai:	a song in praise of God Puma
Pialral:	heaven
Ramhuai:	evil spirit
Ran lu kima ai:	full rounds of animal heads
Sap or saab:	sahib or sir
Sirkar/ sawrkar/ Sarkar:	government
Sachchian:	customary larger share of the hunt for the chief
Singtang:	a variety of rice
Tampuii mithi:	the great famine
Thingtam:	famine resulted by the flowering of the *thing* variety of bamboo
Thangang:	a species of insect
Tlowmngaihna:	philosophy of community before self
Thikadar:	contractor
Upa:	Lushai elders
Vai:	plainsmen
Vai chia:	unscrupulous plainsmen
Vai pa:	nice plainsmen
Yum:	white ants
Zu:	Lushai rice beer

Bibliography

PRIMARY SOURCES

Administrative Report of the Lushai Hills for the years 1881–4, Mizoram State Archives (MSA).

Annual Report of the Baptist Mission Society, South Lushai Hills, Assam, 1912.

Files on *Thingtam* and *Mautam*, DC (A), No. 16/24, 1924–5, MSA.

General Administration Report, Cachar District, Assam Secretariat Press, 1860–1.

Government of Mizoram, *Mautam* (bamboo flowering, rodent outbreaks and famine in Mizoram): The Experiences, Agricultural Extension series no. 1/2806, Aizawl, 2006.

Hennikker, F.C., *Maotam in Lusai: Notes Compiled in July 1912*, Maidstone, 1930, p. 6. Hennikker Papers, Box no. 10, Cambridge: Centre for South Asian Studies, University of Cambridge.

List of Number of Representatives from the Area of Mizo Chiefs, 1946, MSA.

Mackenzie, Alexander, *History of the Relations of Government with Hill Tribes of North Eastern Frontier of Bengal*, Calcutta, 1884, rpt. as the *North East Frontier of India*, Delhi: Mittal, 1994.

'Memorandum of the case of Mizo' Memorandum of the Mizo Union to the Sub-Committee on North-East India excluded and partially excluded areas (Assam) known as 'Bordoloi Committee', MSA.

Memorandum Submitted to the Government of India by Mizo National Council, MSA.

Mizo Union General Assembly Proceedings, General Department Files belonging to the Office of the District Superintendent.

Mizo Union Mite Hnena Thuchah (Memorandum of the Mizo Union), Aizawl, 1947.

Mizo Union Constitution and the Presidential Speech, MSA.

Parliamentary Debates (Official Report), fifth series, vol. LIII, 53, House of Commons of the 30th Parliament, 1913.

Pachhunga Dahrawka and Hmartawnpunga, *Independence*, Aizawl, 29 July 1947.

Pachhunga Dahrawka and Hmartawnpunga, *Mizoram Dindantur Ngaituahhuna Vantlang* (A Public Meeting Dealing with the Future Political State of Mizoram), Aizawl, 1947.

Proceedings of the Foreign Department—1880-1884, vol. I, NAI (Files of Lushai Famine).

Proceedings of the General Department, 10 Famine Reliefs, 1911–12.

Proceedings of the Accredited Leaders of all Lushai Political Parties, Aizawl, 14 August 1947 (in Mizo).

Proceedings of the Assam Legislative Assembly, March 1966.

Proceedings of the Mizo Hills District Council, 1957–9.

Ronghaka, D., *Zoram Independent*, Aizawl, 5 May 1947.

REPORTS AND PAMPHLETS (PUBLISHED)

General Administration Report, 1891–2, Aizawl, MSA.

Reports of the Baptist Mission Society, 1901–38, Aizawl, 1993.

Report on the North-East Frontier (Assam) Tribal and Excluded Areas, 3 vols.

Slavery in British India, Assam and Burma by Peter Fraser, NLA, CMA, Aberystwyth.

'The Bawi System of Lushai, Assam, British India', Letter sent to Colonel Cole, Superintendent of Lushai Hills through Mr. Von Morde, Assistant Superintendent, December 1909, NLA, CMA 27, 318, File V, The Bawi system in Lushai: Dr. Fraser's Case and Letters 1911–13, Aberystwyth, Wales, UK.

The Assam Lushai Hills District (Acquisition of Chiefs Right) Act, 1954.

United Mizo Freedom Party Constitution and its Correspondence with the Superintendent, Lushai Hills, MSA.

Vanlawma, R., *The Mizo Union*, Aizawl, 1946.

Weekly Confidential Reports of the Lushai Hill District for the years 1946, 1947, 1960, 1961, MSA.

Zawla, K., *Zoram Din Hmun Dik Hmuh Chuah Thelhna Tur* (A Guide towards Making the Right Political Decision for Future Mizoram), 7 June 1947.

ACCOUNTS AND MANUSCRIPTS

An Interim Report on the Socio-Economic Survey of Mizo District 1962, Aizawl: Office of the District Statistical Officer, 1963.

Brown, H., *The Lushais, 1888–89*, Shillong: Govt. of Assam, 1890.

Elly, Colonel E.B., *Military Report on the Chin-Lushai Country*, Simla: Government of India, 1893; rpt. Aizawl, Tribal Research Center, 1980, pp. 14–15.

Evidences in the Report on the North East Frontier Tribal and Excluded Areas, vol. 3, by eminent Mizo leaders.

Jones, D E, *A Missionary's Autobiography 1897–1927*, tr. from Welsh by J.M. Llyod, Aizawl: Liansailova, 1988.

Pahlira, C., 'Mizo Hills in the Indian Union', in *Tribal Mirror*, vol. 6, Karimganj, 1970.

Rokhuma, C., *What is the Anti-Famine Campaign Organisation Doing?*, Aizawl: Author, 1988.

Saprawnga, Ch., *Ka Zin Kawng* (My Journey), Aizawl: Author, 1973.

———, 'Factions Contributing to the Mizo Problem', in *Tribal Mirror*, vol. 3, Karimganj, 1967.

Thanlira, R., 'Mizo Union Nawrh' (unpublished) (in English).

Vanlawma, R., *Ka Ram Le Kei* (My Country and I): A Political History of Modern Mizoram, Aizawl: Author, 1972.

———, 'Reminscence of Gopinath Bordoloi', in Lily M. Barua, *Lokpriya Gopinath Bordoloi*, Delhi: Gyan, 1992.

INTERVIEWS

Interview with C. Rokhuma on 22 September 2001 at his residence at Mission Veng. Rokhuma is an activist of the Anti-Famine Organization and an authority on the subject.

Interview with R. Vanlawma on 8 July 1998 at his home Venglui, Aizawl. He was born in 1915. This Octogenarian was the first matriculate among the Mizos. He was the leader of 'Young Lushai Association' which was later renamed as 'Mizo Union'. It may be noted that he was awarded 'Padmashree' on 26 January 1998 by the President of India.

SECONDARY SOURCES

Books and Articles

Agarwal, Anil et al. (eds.), *State of India's Environment: A Citizen's Report*, Delhi: CSE, 1982.

Anderson, Benedict, *Imagined Communities: Reflections on the Origin and Spread of Nationalism*, London: Verso, 1991.

Baker, E.C.S., 'The Game Birds of India Burma and Ceylon', in *Journal of Bombay Natural History Society*, 24, 1916, pp. 201–3.

Bayly, Susan, *Saints, Goddess and Kings: Christians and Muslims in South Indian Society 1700–1900,* Cambridge, 1989.

Barkakaty, S., *Tribes of Assam,* Gauhati, Assam Publication Board, 1969.

Barpujari, H.K., *Problem of the Hill Tribes in North East India*, 3 vols., Gauhati: Lawyers, 1970, 1977 and 1981.

———, *The American Baptist Missionaries in North East India 1836-1900: A Documentary Study*, Gauhati: Spectrum, 1986.

———, *The Comprehensive History of Assam*, vol. IV, Gauhati: Publication Board, 1992.

Barpujari, S.K., 'Bamboo Flowering in Mizoram', in *Proceedings of North East History Session*, Aizawl, 1996.

Baruah, A., *Lokpriya Gopinath Bordoloi*, Delhi: Govt. of India, 1992.

Barooah, Nirode K., *David Scott in North East India: A Study in Paternalism, 1802–31*, Delhi: Munshiram Manoharlal, 1977.

Baveja, J.D., *Mizoram: The Land Where Bamboo Flowers*, Gauhati: Assam Publication Board, 1970.

Bhattacharjee, J.B., *Cachar, Under British Rule in North East India*, Delhi: Radiant, 1977.

Blanford, H.R., 'Note on Operations in Bamboo Flowered Areas in Kartha Division', *Indian Forester*, 44, 1918, pp. 550–60.

Blatter, E., 'The Flowering of the Bamboo (part I)', *Journal of Bombay Natural History Society*, 33, 1929, pp. 899–992.

Bourdillon, T.F., 'Seeding of the Bamboo Theory', *Indian Forester*, 21, 1895, pp. 228-9.

Bower, Ursula Graham, *Naga Path*, London: Readers Union, 1952; rpt. Gauhati: Sepectrum, 2002.

Bradley, J.W., 'Flowering of Kija Thanung Bamboo (Bambusa Polymorpha in Ironme Division, Burma', *Indian Forester,* 25, 1899, pp. 1–25.

Brandis, D., 'Biological Notes on Indian Bamboos', *Indian Forester*, 25, 1899, pp. 25–50.

Butler, John, 'Rough Notes on Angami Naga', *Journal of Asiatic Society of Bengal,* vol. 44, no. 4, 1875, pp. 307–27.

Chakravarty, Ranjan, ed., *Does Environmental History Matter? Shikar, Substance, Sustenance and the Sciences*, Kolkata: Readers Service, 2006.

Chakravarty, Suhas, *The Raj Syndrome: A Study in Imperial Perceptions*, Delhi: Penguin, 1991.

Chaloult, N.B. and Y. Chaloult, 'The Internal Colonialism Concept: Methodolcgical Consideration', in *Social and Economic Studies*, 1979, vol. 28, no. 4.

Chaube, S.K., *Hill Politics in North East India*, Delhi: Orient Longman, 1973; rpt. 1999.

Chatterjee, D., 'Bamboo Fruits', *Journal of Bombay Natural History Society,* 57, 1960, pp. 451–3.

Chatterjee, S., *Mizoram under British Rule*, Delhi: Mittal, 1985.

———, *Mizo Chiefs and Chiefdom*, Delhi, 1995.

———, *Mizoram Encyclopaedia*, 3 vols., Bombay: Jaico, 1990.

———, *A History of the Mizo Economy*, vol. I, Jaipur: Printwell, 1999.

Choudhury, S., *Social Background of Mizo Insurgency*, Shillong: ICSSR, North-Eastern Regional Centre.

Cutter, Susan L., *Living with Risks: The Geography of Technological Hazards*, London, 1993.

———, ed., *Environmental Risk and Hazards*, New Jersey, 1994.

David Raju B., 'The Relief Activities of the Christian Missionaries in South Coastal Andhra During the Famine of 1876–78 and Mass Conversions', paper presented and abstract published in the *Proceedings of the Indian History Congress*, 53rd Session, Warrangal, 1992–3, Delhi, 1993. Abstract no. 96.

Downs, Frederick, *Essays on Christianity in North East India*, ed. Milton S. Sangma and David Syiemlieh, Delhi: Indus, 1994.

Dove, J.L, 'Modeling Internal Colonialism: History and Prospect', *World Development*, 1989, vol. 17, no. 6.

Eaton, Richard E, 'Conversion to Christianity Among the Nagas, 1861-1971', *Indian Economic and Social History Review*, vol. 11, no. 1, 1984,

Frykenberg, Robert Eric (ed.), *Christians and Missionaries in India: Cross Cultural Communications Since 1500*, London: Routledge/Curzon, 2003.

Gait, E., *A History of Assam*, Calcutta: Thacker Spink & Co., 1905.

Gautam, P. K., 'Don't Nip at the Bud', *Down to Earth,* 16 February 2004.

Geddie, Willian, ed., *Chambers Twentieth Century Dictionary*, Edinburgh: W. & R. Chambers, 1964 rpt.

Glickmen, T.S., D. Golding and E.D. Silverman, 'Acts of God and Acts of Man: Recent Trends in Natural Disasters and Major Industrial Accidents', Washington D.C.: Centre for Risk Management, Discussion Paper, CRM, 1992–2000; cited in Susan L. Cutter, ed., *Environmental Risk and Hazards*, New Jersey, 1994.

Goswami, B.B., *Mizo Unrest: A Study of Politicisation of Culture*, Jaipur: Alekh, 1979.

Guha, Amalendu, *Planter Raj to Swaraj: Freedom Struggle and Electoral Politics in Assam 1826–1947*, Delhi: ICHR, 1977.

———, 'Nationalism: Pan Indian and Regional', Presidential Address, Indian History Congress, Modern India Section, 1982.

Hansaria, B.L., *Sixth Schedule to the Indian Constitution: A Study*, Gauhati: Author, 1983.

Hechter, Michael, 'Internal Colonialism Revisited', in E.A. Tirykian and R. Rogowski, eds., *New Nationalisms of the Developed West*, Boston: Allen and Unwin, 1985.

Hind, Robert J., 'The Internal Colonial Concept', in *Comparative Studies and Society and History*, vol. 26, no. 3, 1984.

Hluna, J.V., *Church and Political Upheaval in Mizoram*, Aizawl: Author, 1988.

Hminga, C.L., *The Life and Witness of the Churches in Mizoram*, Aizawl: Baptist Church, 1987.

Hutchinson, Francis G., *The Illusion of Permanence: British Imperialism in India*, Princeton: Princeton University Press, 1967.

International Network for Bamboo and Rattan, http:/www.inbar. int/flowering/Assets/Quiz%20 answers,htm

International Seminar on the Mizos held at Aizawl, 7–9 April 1992, *Proceedings*, Aizawl, 1992.

Janzen, D., 'Why do Bamboos Wait so Long to Flower?' *Annual Review of Ecology and Systematics*, 7, 1976.

Jenkins, G.H. and R.R. Davies, eds., *From Medieval to Modern Wales: Historical Essays in Honour of Kenneth O. Morgan and Ralph A. Griffith*, Cardiff: University of Wales Press, 2004.

Jones, D.E., *A Missionary's Autobiography, 1897–1927*, Aizawl: Liansailova, 1988.

Lalbiakthanga, *The Mizos: A Study in Racial Personality*, Gauhati: United Publishers, 1978.

Ladurie, Emmanuel Le Roy, *Times of Feast, Times of Famine: A History of Climate since the Year 1000*, London, 1972.

Lalchungnunga, *Mizoram: Politics of Regionalism*, Gauhati: United Publishers, 1996.

Laldena, *Christian Missions and Colonialism: A Study of Missionary Movement in North East India with Particular Reference to Manipur and Lushai Hills 1894–1947*, Shillong: Vandrame Institute, 1988.

Laldinpui, 'The Mizo Non-Cooperation Movement', in *North East India History Proceedings*, Aizawl, 1996.

Lalsawma, 'Four Decades of Revivals: The Mizo Way', *A Gospel Centenary Souvenir*, Aizawl: Author: 1994.

Llyod, J.M., *On Every High Hills*, Aizawl: Synod, 1984.

———, *History of the Church in Mizoram* (*Harvest in the Hills*), Aizawl: Synod, 1991.

Lorrain, R.A., *5 Years in Unknown Jungles for God and Empire*, Lakher Pioneer Mission, Liverpool, 1912; rpt., Gauhati: Spectrum, 1988.

Maslekar, A.R., 'More on Bamboo Flowering', *Economic and Political Weekly*, 13–20 December 2003.

Mayhew, Arthur, *Christianity and the Government of India 1600–1920*, London, 1929.

McCosh, John, *Topography of Assam*, London, 1837; rpt., Guwahati: Spectrum, 1986.

McCall, A.G., *Lushai Chrysalis*, London, 1949; rpt., Kolkata: Firma KLM, 2004.

Mclean, Iain, *Concise Dictionary of Politics*, New York: Oxford University Press.

M. Mechter and M. Levi, 'The Comparative Analysis of Ethno-Regional Movements', *Ethnic and Racial Studies*, vol. 2, no. 3, 1979.

Moore, P.H., 'Need of a Native Ministry', in Papers and Discussions of the Jubilee Conference of the American Baptist Missionary Union Held in Nowgong, December 18–29, 1886, Guwahati: Spectrum, 1997.

Morris, J.H., *History of Welsh Calvinistic Methodist Foreign Mission*, Aizawl: Synod, 1910.

———, *The Story of Our Foreign Mission*, Aizawl: Synod, 1930.

Nag, A.K., *The Mizo Dilemma*, Silchar: Tribal Mission Publsihers, 1984.

Nag, C.R., *The Mizo Society in Transition*, Delhi: Vikas, 1994.

Nag, Sajal, *India and North East India: Mind, Politics and the Process of Integration 1946–50*, Delhi: Regency, 1998.

———, 'Rats, Tribal, State and Nation', in *Economic and Political Weekly*, vol. 34, no. 5, 22 December 2001.

Nag, Sajal and Satish Kumar, 'Noble Savage to Gentlemen: Discourses of Civilization and Missionary Modernity in North East India', *Contemporary India*, Journal of Nehru Memorial Museum & Library, New Delhi, vol. 1, no. 4, October–December 2002.

———, *Contesting Marginality: Ethnicity Insurgency and Subnationalism*, Delhi: Manohar, 2002.

Neil, Stephen, *Colonialism and Christian Missions*, London, 1968.

Nibedan, N., *Mizoram: Daggers Brigade*, Delhi: Lancer, 1980.

———, *North East India: The Ethnic Explosion*, Delhi: Lancer, 1981.

Orridge, A., 'Uneven Development and Nationalism: 1/2', *Political Studies*, vol. 29, no. 2. 1981, p. 190.

Parry, N.E., *The Lakhers*, Calcutta, 1931; rpt. Kolkata, Firma KLM, 1988.

Peddappaiah, R.S. and W. Liese, eds., *Recent Advances in Bamboo Research*, Jodhpur: Scientific, 2003.

Potts, Daniel E., *British Baptist Missionaries in India 1837*, Cambridge: The History of Serampore and Its Missions, 1967.

Porter, Andrew, *Religion versus Empire: British Protestant Missionaries and Overseas Expansion 1700–1914*, Manchester and New York: Manchester University Press, 2004.

Prasad, R.N., *Government and Politics in Mizoram, 1947–86*, Delhi: Concept, 1987.

———, *Autonomy Movements in Mizoram*, Delhi: Concept, 1994.

Ray, Animesh, *Mizoram: Dynamics of Change*, Calcutta, 1982.

———, *Mizoram*, Delhi: NBT, 1993.

Rao, V.V., *Century of Government and Politics in North East India*, Delhi: S. Chand & Co. 1976.

———, *Century of Government and Politics in North East India*, vol. III: *Mizoram*, Delhi: S. Chand & Co., 1991.

Reid, Robert, *History of the Frontier Areas Bordering on Assam from 1883–1942*, Shillong: Govt. of Assam, 1942; rpt., Guwahati: Spectrum, 1997.

Rengsi, V. Ruata, 'Pre-Colonial Technology of the Mizos', unpublished Ph.D. thesis, North-Eastern Hill University, Shillong, 1998.

Rezanov, I.A., *Catastrophe in Earth's History*, Moscow: Progress Publishing House, 1980.

Rustomji, Nari, *Enchanted Frontiers*, Delhi: Oxford University Press, 1973.

———, *Imperiled Frontiers*, New Delhi: Oxford University Press, 1989.

Samanta, Arabindo, *Prakritik Viparjoy O Manush* (in Bangla), Calcutta: Deys' Publishing, 2003,

Saikia, Rajen, *Social and Economic History of Assam 1853–1921*, Delhi: Manohar, 2001

Samuelson, R. Sena, *Love Mizoram*, Imphal: Author, 1985.

Savur, Manorama, 'Saving the Eco-System of North East', in *Economic and Political Weekly*, 15–22 November 2003.

Sema, Hokishe, *Emergence of Nagaland*, Delhi: Vikas, 1986.

Seethalakshmi, K.K and M.S. Mukesh Kumar, *Bamboos of India: A Compendium*, Tiruvananthapuram: Kerala Forest Research Institute, 1982.

Sengupta, K.P., *The Christian Missionaries in Bengal, 1793–1833*, Calcutta: K.P. Bagchi, 1971.

Shakespear, J.W., *The Lushai–Kuki Clans*, London, 1921; rpt., Calcutta: Firma KLM, 2002.

Shanmughavel, P., R.S. Peddappaiah and W. Liese, eds., *Recent Advances in Bamboo Research*, Jodhpur: Scientific, 2003.

Sills, David L., ed., *International Encyclopaedia of Social Sciences*, vol. 5, New York: Macmillan and Free Press, 1968.

Singh, Daman, *The Last Frontier: People and Forest in Mizoram*, Delhi: TERI, 1996.

Singh, Y.S., 'Nupilan: Manipur Womens' Agitation', in *Economic and Political Weekly*, vol. II, no. 8, 21 February 1978.

Studdert-Kennedy, Gerald, *Providence and the Raj: Imperial Mission and Missionary Imperialism*, Delhi: Sage, 1998.

Stokes, Eric, *The English Utilitarians and India*, Delhi: Oxford University Press, 1989.

Santapu, H., 'The Flowering of Strobilantnes', in *Journal of Bombay Natural History Society*, 44, 1944, pp. 605–6.

Taylor, A.E., *Famine*, M.S. Stanged University: Food Research Institute, 1947.

Thanzauva, K. (comp), *Report of the Foreign Missions of the Presbyterian Church of Wales on Mizos, 1894–1957*, Aizawl: Synod, 1997.

Vasanthi, *When Bamboo Blossoms*, English tr. Gomathi Narayanam, Gauhati: Spectrum, 1989.

Vanlanchunga, R. (comp), *Report of the Foreign Mission of the Presbyterian Church ofWales in Sylhet (Bangladesh) and Cachar* (*India*), Silchar: Shalom, 2003.

Verghese, C.Z. et al., *A History of the Mizos*, vols. I & II, Delhi: Vikas, 1997.

Vishwanathan, Shiv, *Carnival of Science*, Delhi: Oxford University Press, 1997.

———, 'House of Bamboo', in *The Carnival of Science*, Delhi: Oxford University Press, 1997, pp. 204–5.

White, W.E., 'The Relations of Missionaries to the Government and their Own Mission Councils, in *Indian Witness*, 4 September 1913.

Wolpe Harold, 'The Theory of Internal Colonialism: The South African Case', in Ivan Oxaal and D. Booth, eds., *Beyond the Sociology of Development: Economy and Society in Latin America and Africa, London: Routledge*, 1975.

Vumson, *Zo History*, Aizawl, not dated.

Yonou, Asoso, *The Rising Nagas: A Historical and Political Study,* Delhi: Vivek, 1974.

Yule, H., 'Notes on the Khasi Hills and People', *Journal of Asiatic Society of Bengal*, vol. 12, pt. 2, July–December 1894.

Zairema, *Gods Miracle in Mizoram*, Aizawl: Synod, 1978.

Zaithanga, V.L., 'From Head Hunting to Soul Hunting', Aizawl: Synod, 1981.

Theses and Dissertations

Bhattacharjee, S, 'Man, Calamity History: The History of Earthquakes and its impact on North East India, 1226–1960', unpublished Ph.D. thesis, Assam University, Silchar, 2004.

Pautu, S., 'Separatist Politics in Mizoram', unpublished M.Phil. thesis, North-Eastern Hill University, Shillong, 1982.

Rengsi, V. Ruata, 'Pre-Colonial Technology of the Mizos', unpublished Ph.D. thesis, North-Eastern Hill University, Shillong, 1998.

Valerie, Grossman, 'Preventing Mautam: Participatory Action Research and Phenomenology at Work to Avoid Rat Induced Famine in Mizoram (India)', unpublished Ph.D. thesis in Fielding University, California, 2005.

Index